I0493848

<u>Disclaimer</u>

Book Title: Simulating Occupancy in The NIST Net-Zero Energy Residential Test Facility

Book Author: Farhad Omar; Steven T. Bushby;

Book Abstract: The Net-Zero Energy Residential Test Facility (NZERTF), at the National Institute of Standards and Technology (NIST) in Gaithersburg, Maryland, is a research house that is comparable in size and aesthetic to the houses in the greater Washington DC metro area. The purpose of the NZERTF is to demonstrate the feasibility of achieving net zero energy over the course a year (i.e., energy generated using photovoltaic modules and solar hot water equals energy consumed). The lifestyle choices of the occupants can have a substantial effect on the overall energy consumption of the house. As a laboratory facility, a methodology was needed to simulate this occupants‰ behavior. The occupancy in the NZERTF is emulated by a virtual family of four whose behavior and activities are based recommendations published by the U.S. Department of Energy. In order to attempt to realistically emulate the daily activities of the virtual family it is necessary to replicate their occupancy profiles, water usage, lighting usage, miscellaneous electric plug loads, cooking loads, appliances loads, and sensible and latent loads. This paper discusses the methodology, strategy, and hardware behind emulating the occupancy in the NZERTF.

Citation: NIST TN - 1817

Keywords: Net zero energy house; net zero energy residential test facility; occupancy profile; occupancy schedule; occupancy emulation; daily occupancy emulation schedule

NIST Technical Note 1817

Simulating Occupancy in the NIST Net-Zero Energy Residential Test Facility

Farhad Omar
Steven T. Bushby

National Institute of
Standards and Technology
U.S. Department of Commerce

NIST Technical Note 1817

Simulating Occupancy in the NIST Net-Zero Energy Residential Test Facility

Farhad Omar
Steven T. Bushby
Energy and Environment Division
Engineering Laboratory

November 2013

U.S. Department of Commerce
Penny Pritzker, Secretary

National Institute of Standards and Technology
Patrick D. Gallagher, Under Secretary of Commerce for Standards and Technology and Director

Disclaimer

Certain commercial entities, equipment, or materials may be identified in this document in order to describe an experimental procedure or concept adequately. Such identification is not intended to imply recommendation or endorsement by the National Institute of Standards and Technology, nor is it intended to imply that the entities, materials, or equipment are necessarily the best available for the purpose.

National Institute of Standards and Technology Technical Note 1817
Natl. Inst. Stand. Technol. Tech. Note 1817, 65 pages (November 2013)
CODEN: NTNOEF

Abstract

The Net-Zero Energy Residential Test Facility (NZERTF), at the National Institute of Standards and Technology (NIST) in Gaithersburg, Maryland, is a research house that is comparable in size and aesthetic to the houses in the greater Washington DC metro area. The purpose of the NZERTF is to demonstrate the feasibility of achieving net zero energy over the course a year (i.e., energy generated using photovoltaic modules and a solar hot water system equals energy consumed). The lifestyle choices of the occupants can have a substantial effect on the overall energy consumption of the house. As a laboratory facility, a methodology was needed to simulate the occupants' behavior. The occupancy in the NZERTF is emulated by a virtual family of four whose behavior and activities are based on recommendations published by the U.S. Department of Energy. In order to attempt to realistically emulate the daily activities of the virtual family it is necessary to replicate their occupancy profiles, water usage, lighting usage, miscellaneous electric plug loads, cooking loads, appliances loads, and sensible and latent loads. This paper discusses the methodology, strategy, and hardware behind emulating the occupancy in the NZERTF.

Keywords

Net zero energy house; net zero energy residential test facility; occupancy profile; occupancy schedule; occupancy emulation; daily occupancy emulation schedule

Acknowledgments

The authors wish to thank everyone involved in the Net Zero Residential Test Facility project. Special thanks to Mark Davis, William Healy, Hunter Fanney, Piotr Domanski, Vance Payne, Harrison Skye, Mark Kedzierski, Tania Ullah, Andy Persily, Lisa Ng, Glen Glaeser, John Wamsley, David Yashar, Sandra Heckman, Monyelle Mingo, and Joshua Kneifel of Engineering Laboratory. The authors also wish to thank Michael J. Vanderploey, Vincent P. Anderson, and Alan R. Frank of Whirlpool Corporation, Charlie Smith of GE Appliances, and Benjamin King of BSH Home Appliances.

Authors Information

Farhad Omar
Electrical Engineer
National Institute of Standards and Technology
100 Bureau Drive, Mailstop 8631
Gaithersburg, MD, 20899
Tel: 301-975-4008
Email: farhad.omar@nist.gov

Steven T. Bushby
Mechanical Engineer, Leader of Mechanical Systems & Controls Group
National Institute of Standards and Technology
100 Bureau Drive, Mailstop 8631
Gaithersburg, MD, 20899
Tel: 301-975-5873
Email: steven.bushby@nist.gov

Contents

Abstract iii

Acknowledgments iv

Authors Information iv

List of Figures vi

List of Tables vii

1. Introduction 1

2. Design Basis for Emulating Occupancy 3

    2.1 Family Size 4

    2.2 Occupancy Schedule 4

    2.3 Water Usage 8

    2.4 Lighting Usage 12

    2.5 Miscellaneous Electric Plug Loads 12

    2.6 Cooking Loads 13

    2.7 Appliances Load 13

3. Emulation of the Virtual Occupants 14

    3.1 Sensible and Latent Loads 14

    3.2 Water Usage 17

    3.3 Miscellaneous Plug Loads 20

    3.4 Cooking Loads 23

    3.5 Appliances 25

4. Conclusion 27

References 29

List of Figures

Figure 1. The NZERTF House Exterior 1

Figure 2. The NZERTF first floor plan 2

Figure 3. The NZERTF second floor plan.. 3

Figure 4. Build America benchmark occupancy profile by day type and space type Occ-WD (occupancy week days) and Occ-WE (occupancy weekends) – 4

Figure 5. NZERTF occupancy profile for Mondays and Wednesdays 5

Figure 6. NZERTF occupancy profile for Tuesdays and Thursdays.. 6

Figure 7. NZERTF occupancy profile for Fridays 6

Figure 8. NZERTF occupancy profile for Saturdays 7

Figure 9. The occupancy profile for Sundays.. 7

Figure 10. Median number of draws and variation for households at various levels of occupancy, adopted from [3], Figure 8... 10

Figure 11. Saturday's Lighting Profile.. 12

Figure 12. Latent heat generators 16

Figure 13. Sensible heat generator box and its wiring diagram 17

Figure 14. Water distribution manifold and control solenoids... 18

Figure 15. Master bathroom water measurement apparatus.. 19

Figure 16. Kitchen sink water measurement apparatus.. 19

Figure 17. The heating elements for plug loads with larger than 200 W power requirement and their relay-boxes 20

Figure 18. The relay-box, its wiring diagram, and a toaster plugged into it 21

Figure 19. The electric two-burner portable hot plate, the range hood, and the relay-box 24

Figure 20. The oven Controller-box and its wiring schematic... 25

Figure 21. The dryer, modified misting nozzle, and the washer 26

List of Tables

Table 1. Total volume of water consumption from showers, baths, and sinks for 4 bedrooms (N_{br} = 4). (Equations adapted from [2], Table 10.)* 8

Table 2. The total daily water consumption per person per event and the remaining water volume for sinks use* 9

Table 3. The daily number of kitchen and bathroom sink draws 11

Table 4. The daily distribution of water consumption by the kitchen and bathroom sinks* 11

Table 5. Weekly shower and bath schedule 11

Table 6. Miscellaneous electric plug loads 13

Table 7. Daily cycles for clothes washer, clothes dryer, and dishwasher 14

Table 8. Estimated volume of moisture during cooking events, adapted from [9], Table labeled Household moisture sources* 15

Table 9. Total daily moisture generated by cooking events and the NZERTF occupants 15

Table 10. The plug loads emulation details* 22

Table 11. Cooking appliances emulation details 23

1. Introduction

The Net-Zero Energy Residential Test Facility (NZERTF), at the National Institute of Standards and Technology (NIST) in Gaithersburg, Maryland, is a research house that is comparable in size and aesthetics to the houses in the greater Washington DC metro area. The NZERTF serves two purposes: the first is to demonstrate the feasibility of achieving net zero energy operation – energy generated using photovoltaic modules and solar hot water heaters equals the total energy consumed – over the course of one year. Second, the facility will be used to test existing and new energy efficient and smart grid technologies, and develop test methods that better reflect how those technologies will perform in a real home, rather than in a laboratory. The NZERTF construction was funded by the American Recovery and Reinvestment Act (ARRA). The exterior of the NZERTF is shown in Figure 1.

Figure 1. The NZERTF House Exterior

The NZERTF is a 251 m² (2700 ft²), four bedroom house with a detached garage built entirely with commercially available products. Figure 2 shows the floor plan for the first floor that includes a living room, dining room, kitchen, a bedroom/office, a bathroom, and an unconditioned porch. Figure 3 shows the floor plan for the second floor containing a master bedroom, master bathroom, two other bedrooms, and a bathroom. The detached garage houses the control and measurement equipment used to operate the house. The energy used by the control and measurement equipment and to space condition the garage is not considered part of home's energy consumption.

The activities and habits of the occupants have a major impact on the energy consumption of a home. However, the NZERTF is not occupied by people. Instead the daily activities of the NZERTF occupants are emulated by hardware and software. In the initial year of operation, a virtual family of four is prescribed as the occupants of the facility. The software activates major appliances, plug loads such as entertainment systems, and specially designed devices built to represent the latent and sensible heat generated by the activities of the virtual human occupants.

This report describes in detail the assumptions made about the virtual family's activities and how the resultant energy loads are imposed upon the house in order to provide a realistic approximation of a family living in the facility.

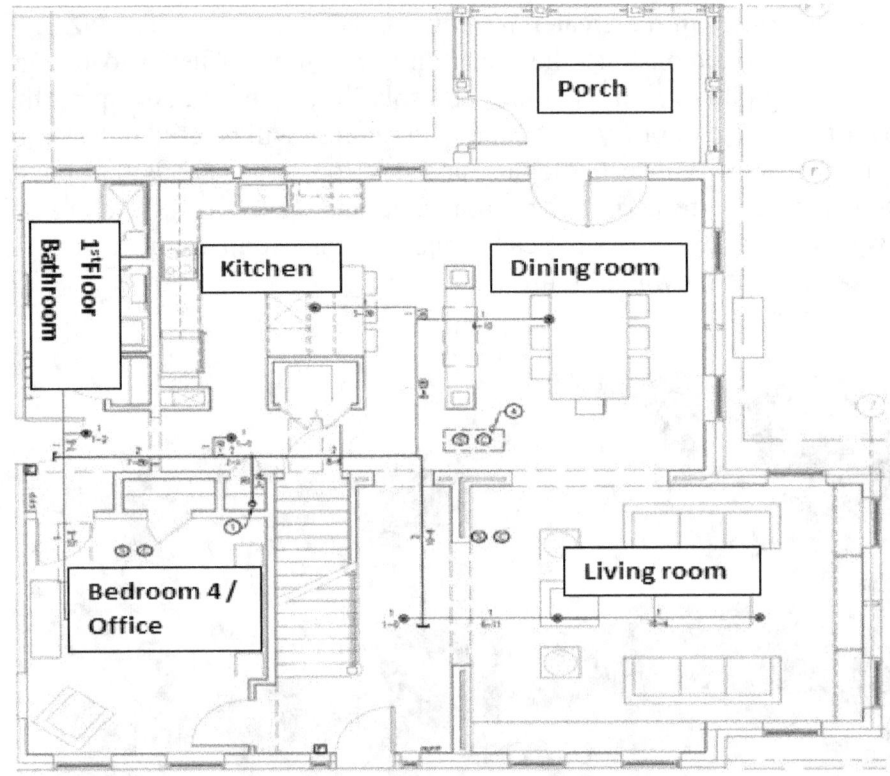

Figure 2. The NZERTF first floor plan

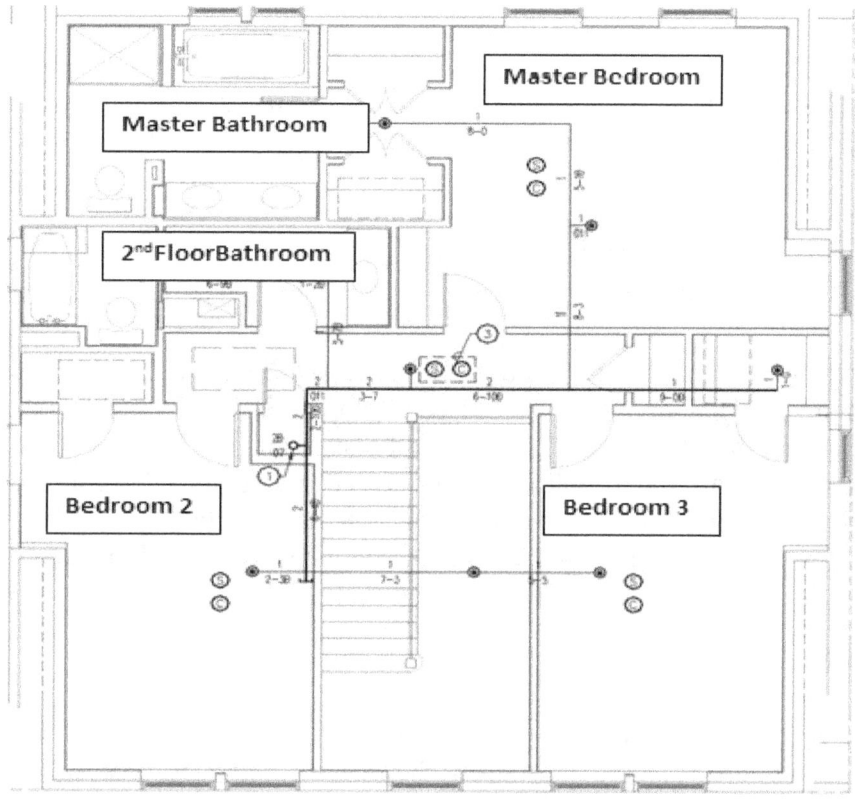

Figure 3. The NZERTF second floor plan

2. Design Basis for Emulating Occupancy

A myriad of detailed choices must be made to realistically emulate the daily activities of a family and the resulting energy impacts. These lifestyle choices can have a substantial effect on energy consumption [1]. Although the NZERTF was designed to have very energy efficient heating, ventilating, and air-conditioning (HVAC) systems, lighting, and appliances, it was assumed that the lifestyle of the occupants would not be substantially different from families living in a conventional house.

Rather than just make an extensive set of arbitrary choices and assumptions about the lifestyle of the virtual NZERTF occupants, the approach taken was to make use of user profiles developed for the U.S. Department of Energy (DOE) Building America program. The Building America program is a multi-year collaboration between DOE national laboratories and top U.S. home builders intended to substantially improve the energy efficiency of homes. In order to measure progress in improving the energy efficiency of homes, the Building America program has established a benchmark intended to be consistent with mid-1990s standard practices [2]. This benchmark includes a series of user profiles intended to represent occupant behavior. The Building America user profiles were used where applicable as the basis for determining all of the details needed to simulate occupancy in the NZERTF.

2.1 Family Size

The first key decision to make is the number of people in the virtual family. According to the survey data reported in [2], the number of occupants in a single-family house can be estimated by Equation 1.

Number of occupants = 0.59 x N_{br} + 0.87	(1)
Where, N_{br} is the number of bedrooms.	

Applying this equation to the four bedroom NZERTF would indicate that the average number of occupants over a large population would be 3.23 people. Since we cannot have fractions of people it was decided that the virtual family would consist of two adults and two children. The ages of the children are arbitrarily selected to be fourteen (middle school age) and eight years old (elementary school age). School attendance factors into the family's daily routines. In this report, the fourteen year old is identified as ChildA and the eight year old is identified as ChildB.

2.2 Occupancy Schedule

The occupancy schedule drives many of the energy loads in the house. To determine the schedule for appliance usage and to account for the sensible and latent load from the people, it is necessary to determine when the occupants are home and when they are in various parts of the house.

Figure 4 shows normalized occupancy patterns for weekdays and weekends that are used in the Building America benchmark [2]. Two space types are considered in this profile; bedrooms and living room.

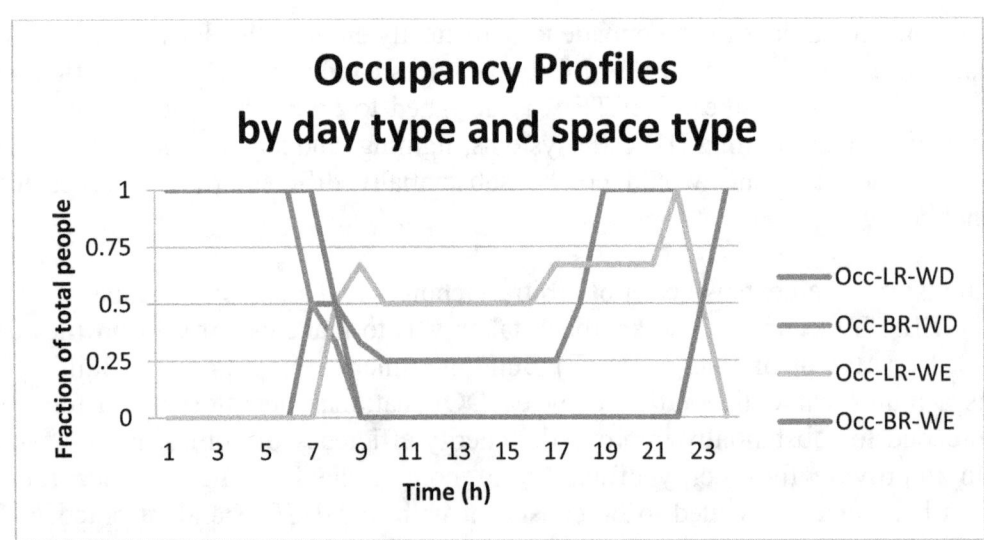

Figure 4. Build America benchmark occupancy profile by day type and space type
Occ-WD (occupancy week days) and Occ-WE (occupancy weekends) –
LR (living room) – BR (bedroom), Adapted from [1] Figure 24.

The occupancy profile, in Figure 4, was used as a guide to develop the occupancy profile details for the NZERTF. Additional details about the virtual family must be assumed to help fill in the daily routine details. It is assumed that both parents work outside the home leaving the house at 8:30 a.m. and returning home at 6:00 p.m. As previously noted, ChildA is a fourteen year old middle school student and ChildB is an eight year old elementary school student. It is assumed that both the middle school and the elementary school start at 8:30 a.m. and end at 3:30 p.m. Arrangements have been made for after school care for ChildB every week day. ChildA has a variable schedule that depends on the day of the week. Both children leave the house at 8:00 a.m. ChildB returns home every day at 6:00 p.m. ChildA returns home at 6:00 p.m. on Mondays, Wednesdays and Fridays, and 4:00 p.m. on Tuesdays and Thursdays. During the summer, it is assumed that the children attend day care or camp outside of the home and, thus, follow the same schedule.

The net result is that on Mondays, Wednesdays, and Fridays the house is unoccupied from 8:30 a.m. to 6:00 p.m. On Tuesdays and Thursdays, the house is unoccupied from 8:30 a.m. to 4:00 p.m. when ChildA returns home. On weekends it was assumed that at least two members of the family are at home at any given time. The family does not take vacations or host large parties such as birthday gatherings.

Figures 5 through Figure 9 represent the daily occupancy profiles of the NZERTF developed by applying these additional constraints to the profile shown in Figure 4. For the NZERTF, occupancy of the living room is taken to mean occupancy anywhere that is not in a bedroom. For simplicity and repeatability, the same weekly schedule is used for 52 weeks, irrespective of the seasons or holidays.

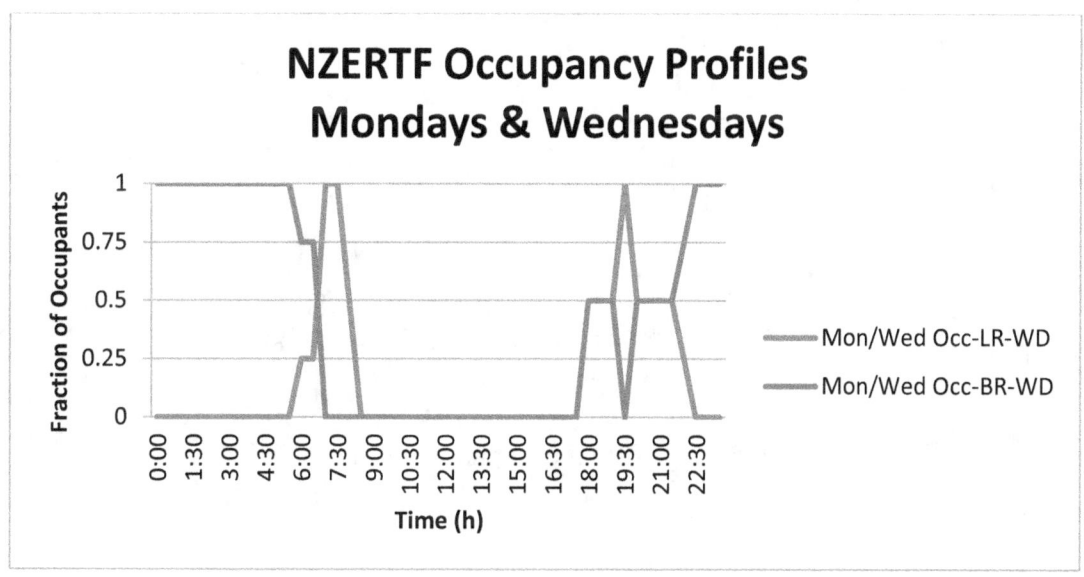

Figure 5. NZERTF occupancy profile for Mondays and Wednesdays

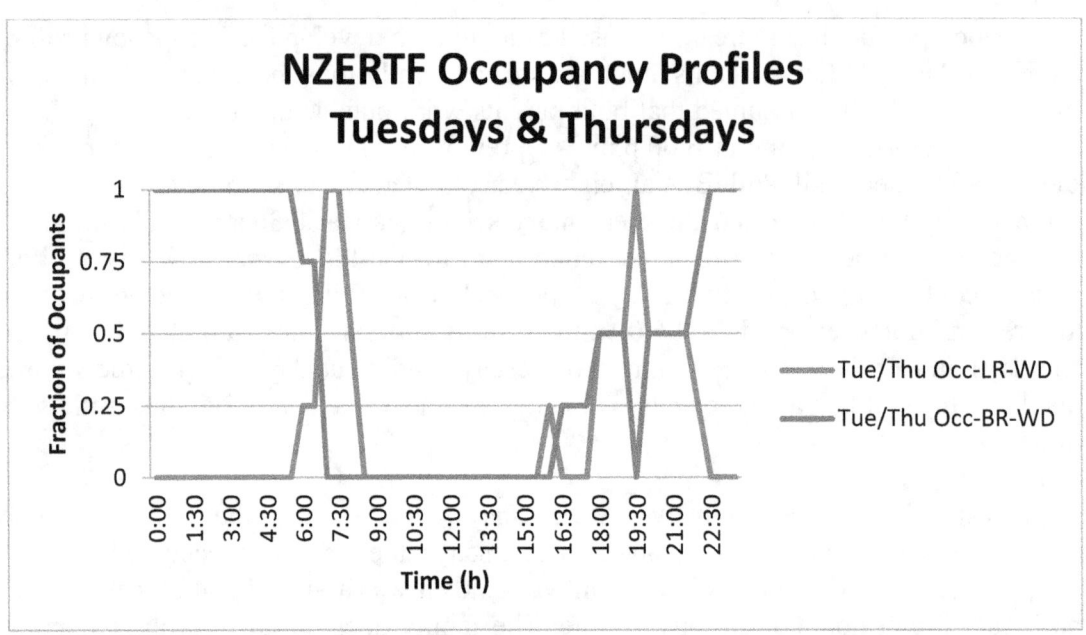

Figure 6. NZERTF occupancy profile for Tuesdays and Thursdays

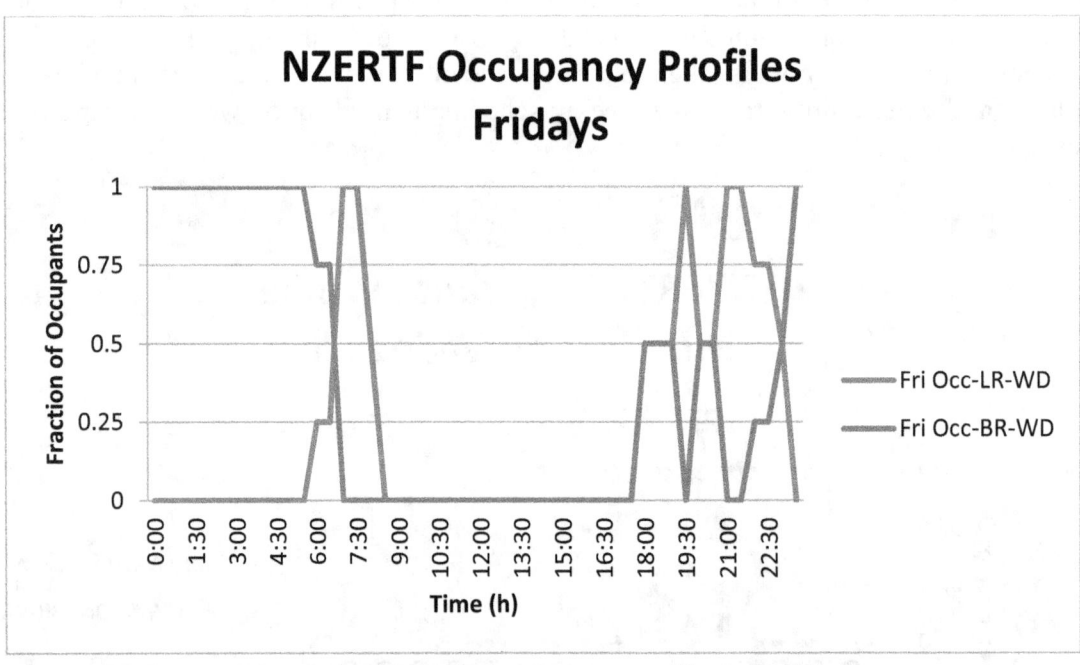

Figure 7. NZERTF occupancy profile for Fridays

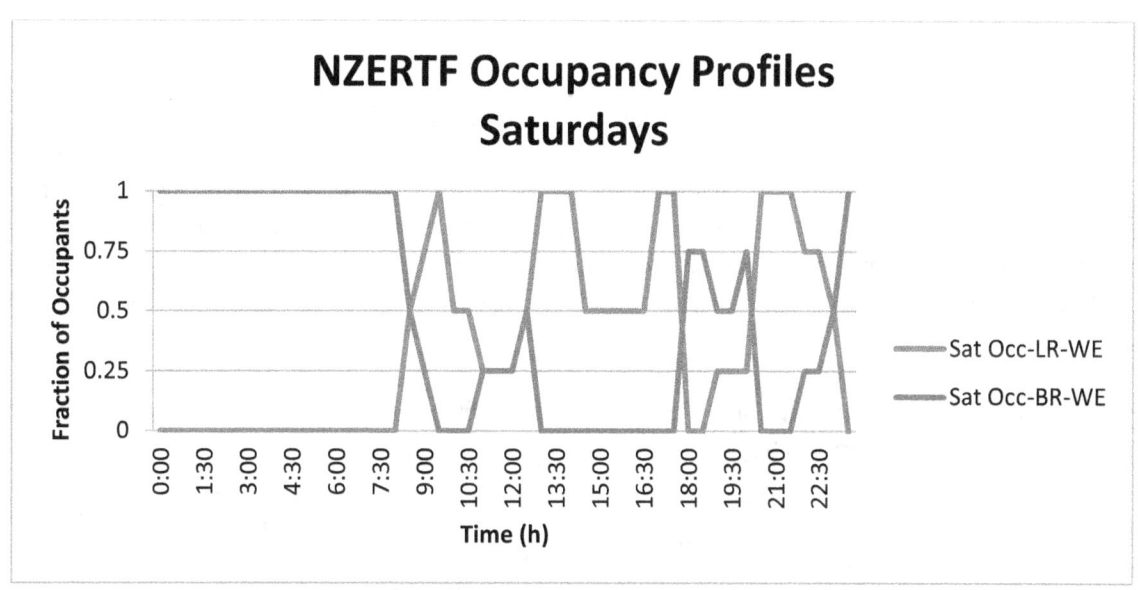

Figure 8. NZERTF occupancy profile for Saturdays

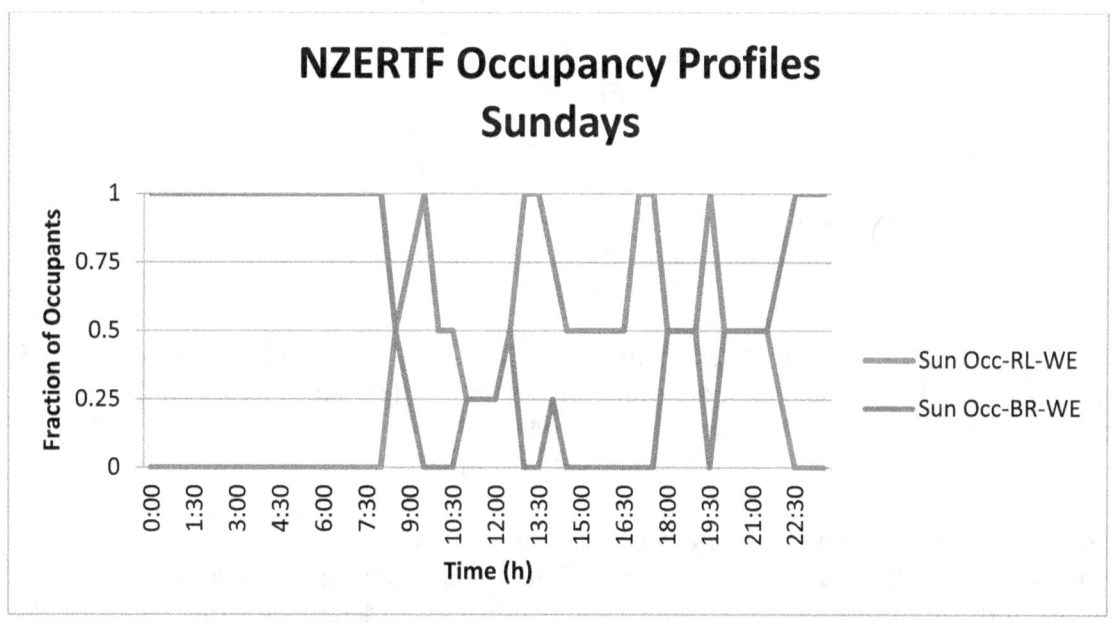

Figure 9. The occupancy profile for Sundays

Establishing the house occupancy pattern enables the development of detailed schedules for other energy consuming activities because occupant initiated activities must take place when the house is occupied.

2.3 Water Usage

Water consumption is an important aspect of the daily activities of the NZERTF occupants. To simulate the water draw activities, it was necessary to identify the total daily volume of water use and identify the mechanisms (i.e., shower and sink use) by which the water is used. As the focus of the facility is on energy use, those end uses that use purely cold water (e.g., toilets) are not simulated. Table 1 summarizes the average daily water consumption from showers, baths, and sinks for a family of four was calculated from the average of three domestic hot-water studies reported in [2].

Table 1. Total volume of water consumption from showers, baths, and sinks for 4 bedrooms (N_{br} = 4). (Equations adapted from [2], Table 10.)*

End Use	End-Use Water Temp.	Water Usage (gallons/day)	Daily Water Usage for Four Bedrooms (liters/gallons)
Shower	40.6 °C (105 °F)	14.0 + 4.67 x N_{br} (Hot + Cold)	123.7 / 32.7
Bath	40.6 °C (105 °F)	3.5 + 1.17 x N_{br} (Hot + Cold)	31.0 / 8.2
Sinks	40.6 °C (105 °F)	12.5 + 4.16 x N_{br} (Hot + Cold)	110.3 / 29.1
Total Daily Volume			**265.0 / 70.0**

*The numbers are calculations used for target set-point and do not reflect the measurement accuracy

According to Table 1, the total daily water usage for showers, baths, and sinks is 265 liters (70 gallons); this number does not include the water consumption by the dishwasher or clothes washer. The dishwasher and clothes washer water usage are determined by their internal mechanisms and we do not restrict their normal operation.

To develop the water draw schedule, we assumed that each occupant bathes once per day. Both parents and ChildA always shower, while ChildB sometimes showers and sometimes takes a bath. Even though some daily variation in bathing was considered desirable, we determined that the weekly volume of water usage must comply with the 1855 liters/week (490 gallons/week) derived from Table 1. Table 2 shows the details of the bathing water consumption assumed throughout the week in order to meet these assumptions.

Table 2. The total daily water consumption per person per event and the remaining water volume for sinks use*

Days	ParentA + ParentB Showers (liters/gallons)	ChildA Showers (liters/gallons)	ChildB Showers or Baths (liters/gallons)	Total Showers & Baths (liters/gallons)	Remaining for Sink (liters/gallons)	Total Daily Volume (liters/gallons)
Monday	66.24 / 17.50	53.00 / 14.00	33.12 / 8.75	152.36 / 40.25	95.32 / 25.18	247.68 / 65.43
Tuesday	66.24 / 17.50	53.00 / 14.00	33.12 / 8.75	152.36 / 40.25	95.32 / 25.18	247.68 / 65.43
Wednesday	66.24 / 17.50	33.12 / 8.75	113.56 / 30.00	212.93 / 56.25	95.32 / 25.18	308.25 / 81.43
Thursday	66.24 / 17.50	53.00 / 14.00	33.12 / 8.75	152.36 / 40.25	95.32 / 25.18	247.68 / 65.43
Friday	66.24 / 17.50	53.00 / 14.00	33.12 / 8.75	152.36 / 40.25	95.32 / 25.18	247.68 / 65.43
Saturday	66.24 / 17.50	33.12 / 8.75	113.56 / 30.00	212.93 / 56.25	95.32 / 25.18	308.25 / 81.43
Sunday	66.24 / 17.50	53.00 / 14.00	33.12 / 8.75	152.36 / 40.25	95.32 / 25.18	247.68 / 65.43

* The numbers are calculations used for target set-points and do not reflect the measurement accuracy

The column labeled as "Remaining for Sink" shows the total daily volume of water to be consumed by sink draws. The total weekly volume of water consumption is 1855 liters (490 gallons), of which 1188 liters (314 gallons) are allocated for showers and baths. The remaining volume of 667 liters (176 gallons) is allocated for the sink draws. The weekly sink volume (667 liters) was equally divided over seven days to compute the daily water consumption for sink events. Recall that the weekly volume of 1855 liters does not include the water consumption by the dishwasher or clothes washer.

The NZERTF sink draws are separated into two categories, bathrooms' sink draws and kitchen sink draws. According to Hot Water Draw Patterns in Single-Family Houses: Findings from Field Studies, the median number of hot water draws for most households ranged between 25 to 100 draws [3].The report in [3] is the result of work sponsored by the California Energy Commission and describes hot water draw patterns that the Lawrence Berkeley National Laboratory obtained from ten studies. Figure 10 shows the variation of median number of daily hot water draws by the number of occupant.

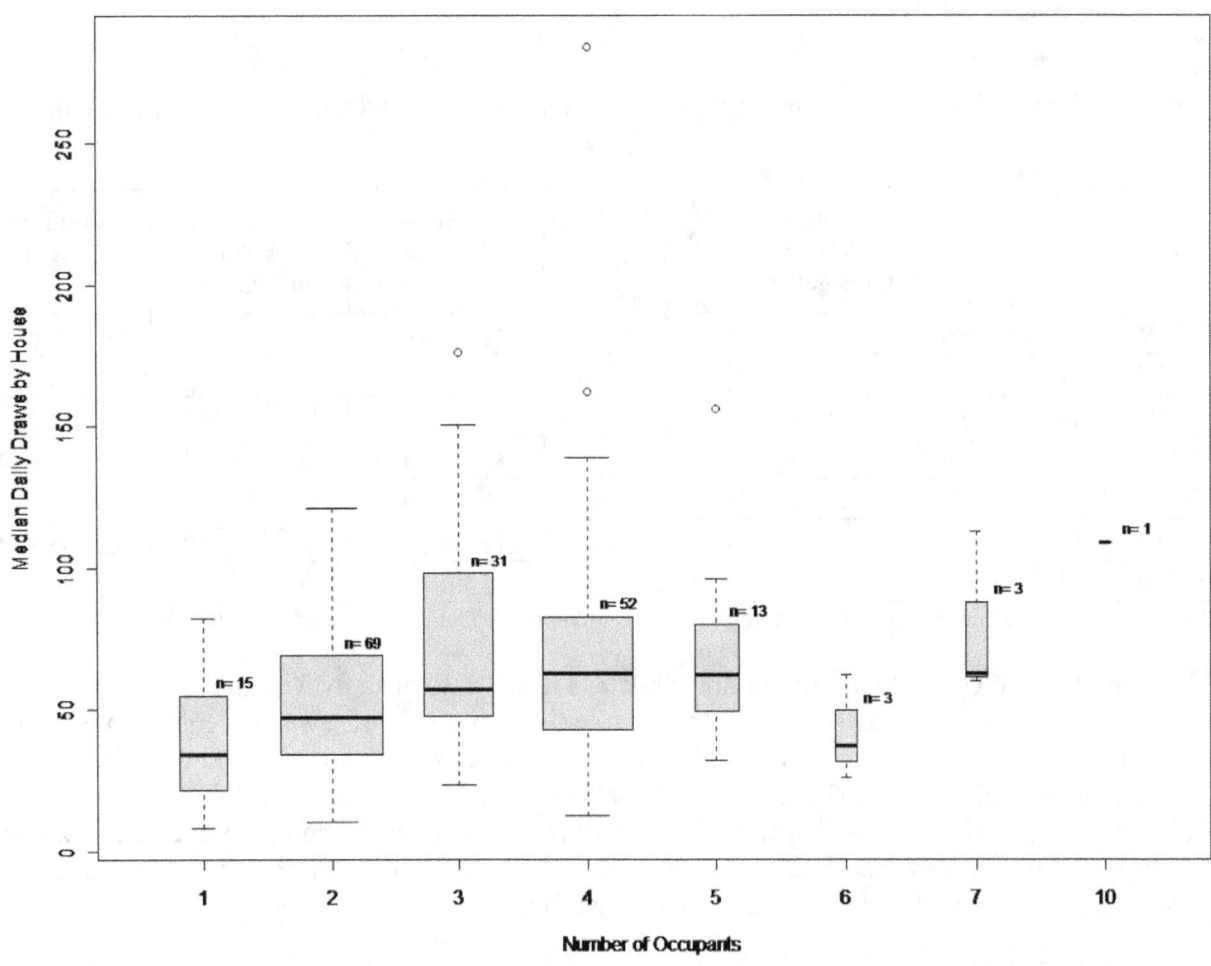

Figure 10. Median number of draws and variation for households at various levels of occupancy, adopted from [3], Figure 8.

In the NZERTF, the goal was to obtain a representative number of water draws that remain within the median range described above. A representative number of draws for the kitchen and bathroom sinks, was obtained from the Standardized Domestic Hot Water (DHW) Event Schedules for Residential Buildings developed by the National Renewable Energy Laboratory (NREL) [4]. The standardized event schedule generator, in spreadsheet format, provides a detailed daily schedule of DHW usage for a year. The methodology used in [4] takes into account the fact that the water main temperature effects the mixture of hot and cold water used. The number of draws for sinks, based on a user-selected typical meteorological year (TMY) site, can be extracted from the schedule. In this case, the TMY for Baltimore-Washington International Airport in Maryland was selected According to the event generator, there are 10 367 kitchen sink and 4439 bathroom sink draws [4]. To obtain the total daily number of draws, the total kitchen and bathroom sink draws were divided by 365 days. Table 3 shows the number of kitchen and bathroom sink draws obtained from the event generator.

Table 3. The daily number of kitchen and bathroom sink draws

Number of Kitchen Sink Draws/Year	10 367
Daily Kitchen Sink Draws	28
Number of Bathroom Sinks Draws/Year	4439
Daily Bathroom Sink Draws	12

The number of draws shown in Table 3 combined with the total sink draw volumes shown in Table 2 was used to compute the volume of water drawn by each sink event. Table 4 shows the daily water consumption by the bathroom and kitchen sink fixtures.

Table 4. The daily distribution of water consumption by the kitchen and bathroom sinks*

Days	Remaining for Sinks (liters/gallons)	Number of Draws Kitchen + Bathroom Sinks	Volume/Draw (liters/gallons)	Total Volume Drawn by the Kitchen Sink (liters/gallons)	Total Volume Drawn by the Bathroom Sinks (liters/gallons)
Monday	95.32 / 25.18	40	2.38 / 0.63	66.72 / 17.63	28.59 / 7.55
Tuesday	95.32 / 25.18	40	2.38 / 0.63	66.72 / 17.63	28.59 / 7.55
Wednesday	95.32 / 25.18	40	2.38 / 0.63	66.72 / 17.63	28.59 / 7.55
Thursday	95.32 / 25.18	40	2.38 / 0.63	66.72 / 17.63	28.59 / 7.55
Friday	95.32 / 25.18	40	2.38 / 0.63	66.72 / 17.63	28.59 / 7.55
Saturday	95.32 / 25.18	40	2.38 / 0.63	66.72 / 17.63	28.59 / 7.55
Sunday	95.32 / 25.18	40	2.38 / 0.63	66.72 / 17.63	28.59 / 7.55

* The numbers are calculations used for target set-points and do not reflect the measurement accuracy

To create the NZERTF water consumption schedule we also needed to identify the time at which the shower and bath events are to take place. The a.m. and p.m. shower/bath schedules follow the occupancy profile discussed in Section 2.2. Table 5 shows the distribution of bathing events.

Table 5. Weekly shower and bath schedule

Days	ParentA / Event Time	ParentB / Event Time	ChildA / Event Time	ChildB / Event Time
Monday	Shower (a.m.)	Shower (a.m.)	Shower (E-p.m.)[1]	Shower (a.m.)
Tuesday	Shower (a.m.)	Shower (a.m.)	Shower (L-p.m.)	Shower (a.m.)
Wednesday	Shower (a.m.)	Shower (a.m.)	Shower (E-p.m.)	Bath (p.m.)
Thursday	Shower (a.m.)	Shower (a.m.)	Shower (L-p.m.)	Shower (a.m.)
Friday	Shower (a.m.)	Shower (a.m.)	Shower (E-p.m.)	Shower (a.m.)
Saturday	Shower (L-p.m.)[2]	Shower (a.m.)	Shower (a.m.)	Bath (p.m.)
Sunday	Shower (p.m.)	Shower (a.m.)	Shower (E-p.m.)	Shower (a.m.)

[1]E-p.m. means early p.m., [2]L-p.m. means late p.m.

A more detailed schedule, that contains water draw events, is provided in the Appendix A of this document.

2.4 Lighting Usage

The NZERTF lighting usage follows the occupancy profile of the family and is irrespective of the seasons' daylight changes. In the evening and morning when someone occupies a space (e.g., a bedroom, living room, or the kitchen) the lights for that space are energized. No lights are energized when the family is sleeping at night except for bathroom events. Similarly, no lights are energized when the house is unoccupied during the week. On the weekends, despite having at least two family members at home at all times, the lights are not energized during the day except for bathroom events. As an example, the NZERTF lighting profile, for Saturday, is shown in Figure 11.

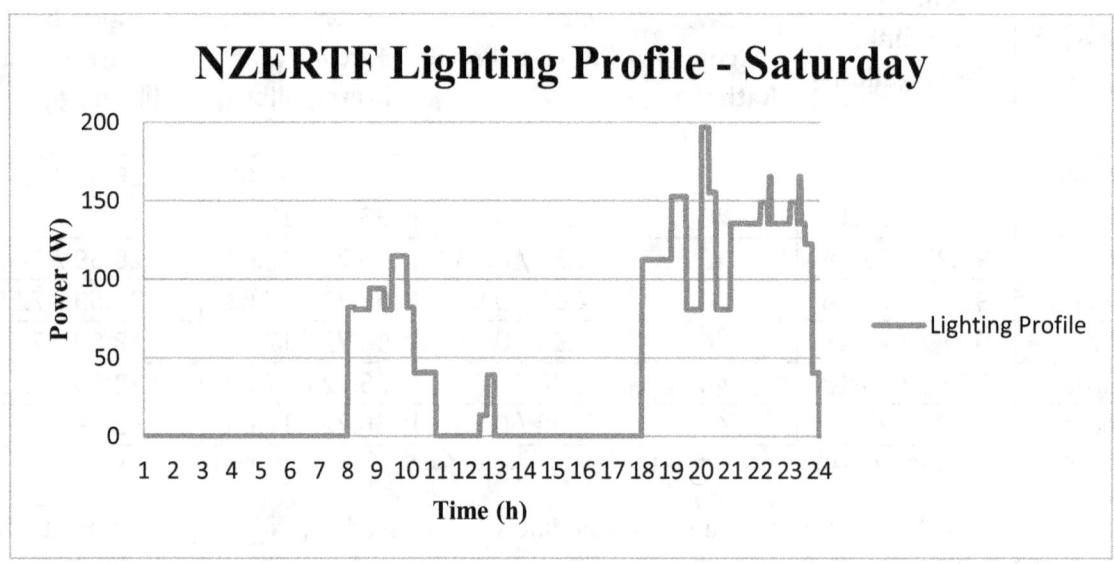

Figure 11. Saturday's Lighting Profile

2.5 Miscellaneous Electric Plug Loads

Simulating occupancy in the NZERTF requires operating various electric plug loads as well as cooking meals and performing laundry. The list of miscellaneous plug loads and their annual energy consumption, unless noted, is obtained from Table 17 in [2]. The criterion to select miscellaneous electric plug loads was assumed to be based on commonly used or owned by at least 50% of households. It is assumed that the annual energy consumption contains the active and standby energy usage. Table 6 shows the list of plug loads and their annual energy consumption.

Table 6. Miscellaneous electric plug loads

Miscellaneous Electric Plug Loads			
Loads	Annual Energy (kWh)	Loads	Annual Energy (kWh)
First Color TV	127.3[1]	Laptop PC	72[2]
Second Color TV	22.3[1]	Desktop PC	237[2]
DVD Player (Blu-ray)	78[2]	PC Monitor	85[2]
Video Game System	41[2]	Printer (inkjet)	39.0
Clock Radio	14.9	DSL/Cable Modem	17.6
Boom Box/ Portable Stereo	16.8	Hair Dryer	41.1
Component / Rack Stereo	153.0	Curling Iron	1.0
Cable Box	152.7	Vacuum Cleaner	55[2]
Microwave	135.1	Clock	26.0
Coffee Maker (Drip)	61.2	Cordless Phone	23.2
Toaster Oven	32.3	Cell Phone Charger	77.4
Toaster	45.9	Answering Machine	33.5
Blender	7.0	Fan (Portable)	11.3
Can Opener	3.0	Heating Pads	3.0
Hand Mixer	2.0	Iron	52.7
Slow Cooker / Crock Pot	16.0	OTHER	9.4
Wireless Router	209.7[3]		

[1] The annual energy consumptions for the First and Second color televisions are based on the occupancy profile. The First one is a 47" LED television and the Second one is a 32" LED television, and both are Energy Star rated

[2] The annual energy consumption is obtained from Table 2.1.16 Operating Characteristics of Electric Appliances in the Residential Sector [5]

[3] The annual energy consumption is obtained by multiplying the estimated power (24 W) by the number of hours in 52 weeks (8736 hours)

2.6 Cooking Loads

Cooking meals are emulated by an electric cooktop and an electric oven. Their annual energy consumption is obtained from Table ES.2 Energy Consumption for Cooking Products [6]. The baseline annual energy for a smooth electric cooktop is 233.4 kWh, and for a self-cleaning electric oven is 303.7 kWh. It is assumed that every time that the cooktop is used the range hood is also operated for the same length of time.

2.7 Appliances Load

With the exception of the HVAC, heat pump water heater, and refrigerator units, the operation of other appliances (e.g., clothes washer, clothes dryer, and dishwasher) are automated by the NZERTF's data acquisition/controller unit. The number of cycles/week for the dishwasher, clothes washer, and clothes dryer was obtained from [7] and is based on a survey conducted by Proctor and Gamble. Table 7 summarizes the daily cycles of clothes washer, clothes dryer, and dishwasher

Table 7. Daily cycles for clothes washer, clothes dryer, and dishwasher

Days	Clothes Washer	Clothes Dryer	Dishwasher
Monday	0	0	1
Tuesday	0	0	0
Wednesday	2	1	1
Thursday	0	0	0
Friday	0	0	1
Saturday	2	2	1
Sunday	2	2	1

The family's interaction with the refrigerator is emulated by placing a thermal load inside the unit. A detailed analysis of this interaction is given in Section 3.5.

3. Emulation of the Virtual Occupants

The discussion of the emulation of the virtual family is separated into several major sections: the implementation of the sensible and latent loads, water usage, miscellaneous plug loads, cooking, laundry, and dishwashing appliances.

3.1 Sensible and Latent Loads

Emulating human occupancy, in a virtual environment, such as the NZERTF, requires accounting for the sensible and latent loads generated by the presence and activities of the occupants themselves. In the NZERTF, sensible loads are simulated by resistors placed in the bedrooms, kitchen, and the living room areas. Each sensible load emulator, representing a particular family member, is activated according to the schedule for that person. For example, when ParentA is in the master bedroom the sensible load emulator for ParentA is activated in the master bedroom. The same procedure is applied to all family members. In contrast, the latent loads produced by the entire family and cooking activities, which are introduced as vaporized water, are lumped together to simplify the instrumentation. This approach can be justified because the NZERTF air-handler unit recirculates the moisture, generated locally in the kitchen, to the entire house. According to Table 1 in Nonresidential Cooling and Heating Load Calculations of ASHRAE's Handbook of Fundamentals, the adjusted sensible and latent load per person, for seated and very light work in an apartment, is 70 W and 45 W, respectively [8]. These adjusted values are averaged for adults and children, so a single value of 70 W is used for all emulators.

To emulate latent load, in the NZERTF, it is essential to determine the amount of moisture dissipated by the family through respiration, perspiration, and cooking activities. According to Latent Heat of Vaporization in Thermal Physiology, at a skin temperature of 33 °C (91.4 °F) the heat needed to vaporize 1 gram of water is 2.398 kJ [9]. The value of latent heat of vaporization was used to convert 45 W to its equivalent volume in liters. The estimated moisture generated

14

per person is 0.07 liters/h (0.02 gallons/h). The latent load in [8] does not include the volume of moisture generated by cooking activities. It is obtained from the Home Moisture document published by the Minnesota Department of Commerce [10]. The estimated latent load generated for each cooking event (denoted as breakfast, lunch, and dinner) is shown in Table 8.

Table 8. Estimated volume of moisture during cooking events, adapted from [9], Table labeled Household moisture sources*

Cooking Event (Family of Four, Average)	Estimated Amount per Event (liters/gallons)
Breakfast	0.17 / 0.04
Lunch	0.25 / 0.07
Dinner	0.58 / 0.15

* The numbers are calculations used for target set-points and do not reflect
  the measurement accuracy

During the week, Monday through Friday, no one is home to prepare lunch so the moisture generated by cooking events only includes breakfast and dinner. The moisture generated on the weekends, however, includes all three meals. The daily moisture generated by the NZERTF family members account for the length of time that they are home. Table 9 shows the daily moisture generated by cooking meals and the occupants.

Table 9. Total daily moisture generated by cooking events and the NZERTF occupants

Days	Daily Moisture from Cooking (liters/gallons)	Daily Moisture from People (liters/gallons)	Total Daily Moisture Cooking + People (liters/gallons)
Sunday	0.99 / 0.26	5.83 / 1.54	6.83 / 1.80
Monday	0.74 / 0.20	3.87 / 1.02	4.61 / 1.22
Tuesday	0.74 / 0.20	4.0 / 1.06	4.74 / 1.25
Wednesday	0.74 / 0.20	3.87 / 1.02	4.61 / 1.22
Thursday	0.74 / 0.20	4.0 / 1.06	4.74 / 1.25
Friday	0.74 / 0.20	3.87 / 1.02	4.61 / 1.22
Saturday	0.99 / 0.26	5.63 / 1.49	6.22 / 1.75

* The numbers are calculations used for target set-points and do not reflect the measurement accuracy

In order to introduce moisture into the NZERTF to represent the latent loads from cooking and the simulated human occupants, an ultrasonic humidifier is used. The approach used and the design of the apparatus are based on recommendations from Field Test Protocol: Standard Internal Load Generation for Unoccupied Test Homes [11]. The advantage of an ultrasonic humidifier is that it creates very fine water droplets as cool fog, thereby releasing a steady rate of moisture at room temperature regardless of the ambient relative humidity.

The apparatus used in the NZERTF, shown in Figure 12, are commercially available humidifiers modified to enable long term operation without human intervention. The water reservoir is connected to a metered water supply line through a solenoid valve operated by a level switch. The solenoid valve is normally closed and the entire device is placed in a pan with a leak detector to enable remote detection of any water leaks.

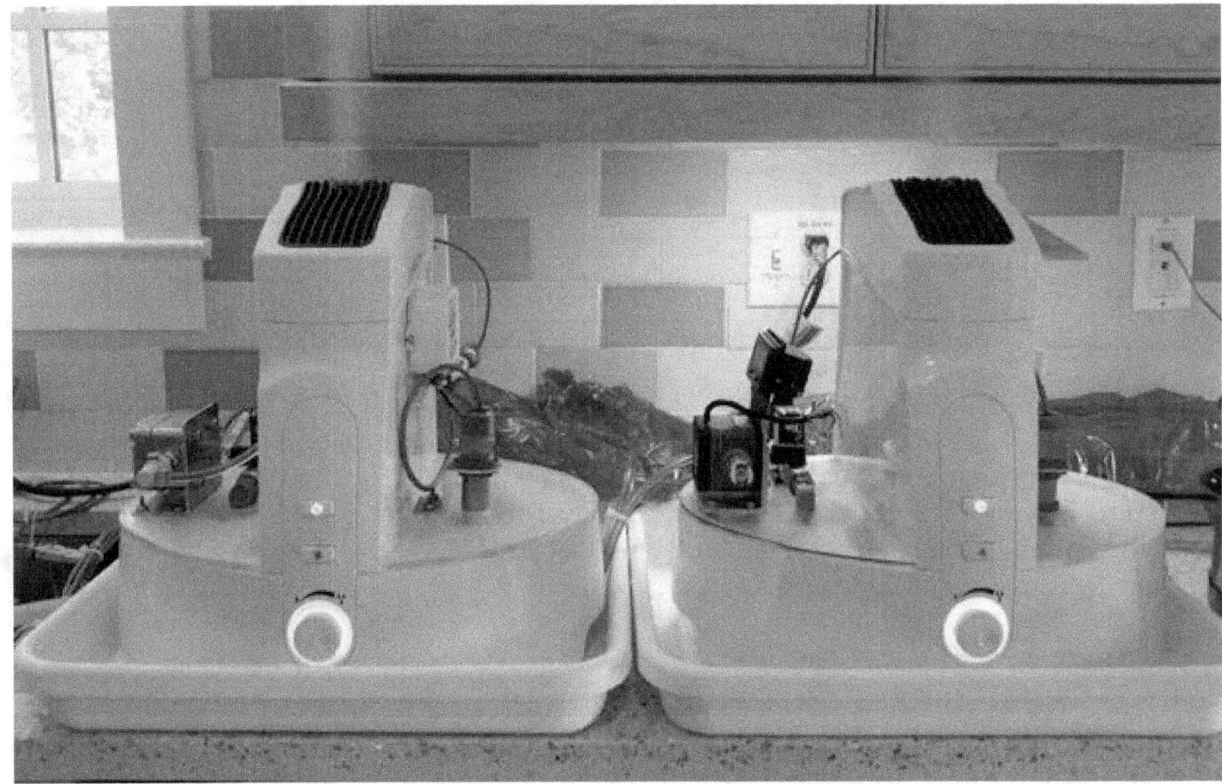

Figure 12. Latent heat generators

Laboratory tests determined that each ultrasonic humidifier's output capacity is approximately 0.27 liters/h (0.07 gallons/h) of moisture. Given this capacity, two ultrasonic humidifiers were installed to meet the daily required latent load as reported in Table 7. The amount of moisture introduced into the air is controlled by time. The flow meter provides a verification of correct operation.

To emulate sensible heat, eight resistance heating boxes were designed and built. Each box represents a family member and placed in bedrooms, kitchen, living room, and dining room. The key components of a box are a bulb socket screw-in heater, a dimmer switch, a solid state relay, a safety fuse, and a fixture to mount the screw-in heater (maximum power rating is 200 W). The dimmer switch controls the output power of each box, which was calibrated to 70 W, to represent the sensible heat generated by each person. The solid state relay turns the sensible heat generator on or off upon receiving a signal from the facility's acquisition/controller unit. The safety fuse is installed in-line with the hot power line to protect the circuit against current overload. Figure 13 shows a sensible heat generator box and the wiring diagram.

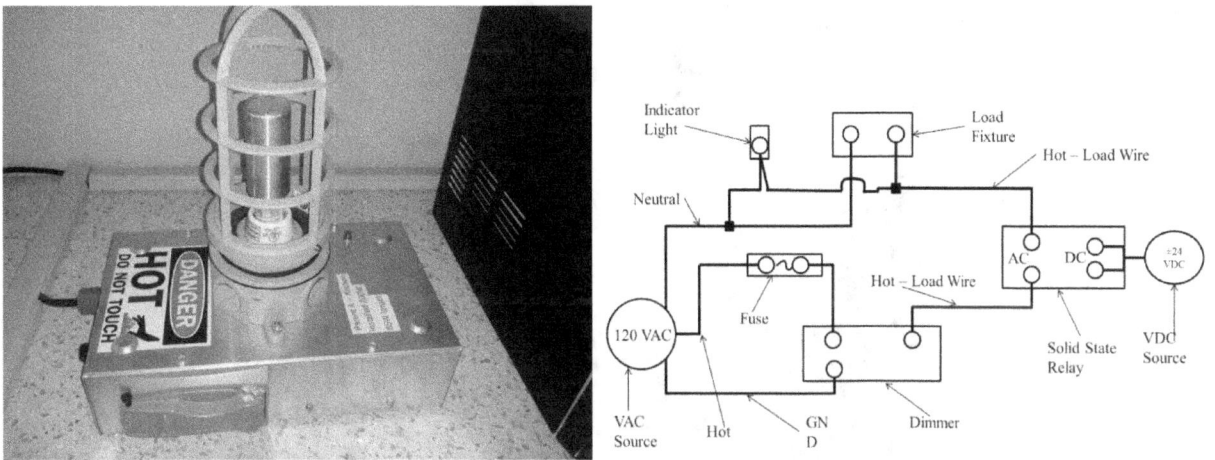

Figure 13. Sensible heat generator box and its wiring diagram

3.2 Water Usage

For simplicity of measurement, all first floor water draws are scheduled in the kitchen sink. No one takes a shower on the first floor, so all first floor events are bathroom and kitchen sink related. All second floor water draws, which include second bathroom and master bathroom shower and sink events, are scheduled in the master bathroom shower. The master bathroom shower is used to emulate the shower and sink events. The bathtub faucet is used to emulate the bath events. While it is acknowledged that this arrangement may not fully capture heat losses in the hot water distribution pipes, the quest to simplify the automation of water draws took precedence over any such inaccuracies.

In the NZERTF, the plumbing distribution system consists of a manifold in the basement from which cross-linked polyethylene tubing delivers hot and cold water directly to each fixture. Normally-closed solenoid valves are installed in each of these lines at the manifold, and a signal is sent to the valves on the hot and cold lines for each fixture to initiate or terminate a water draw. Figure 14 shows the manifold that delivers hot and cold water to each fixture and the control solenoids.

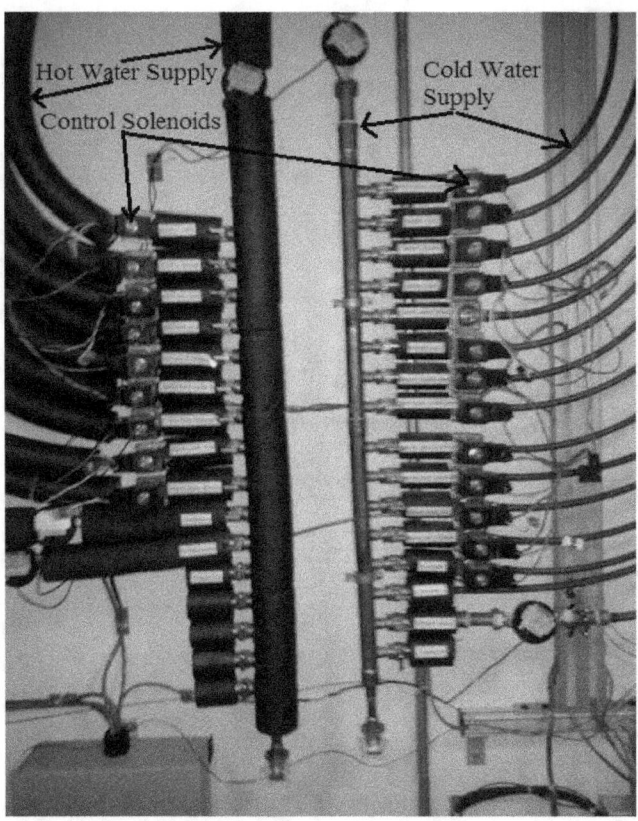

Figure 14. Water distribution manifold and control solenoids

The water from the master bathroom shower and the bathtub faucets are deposited through hoses into a tank placed on a load cell. Similarly, the kitchen sink water is deposited into a smaller tank which is also placed on top of a load cell. At the bottom of each tank, there is a drain valve that is controlled by a normally open solenoid. When a draw event is activated, the solenoid is closed to allow water accumulation in the tank. The scale measures the mass of water accumulated in the tank, with the volume removed being computed based on the density of water at the temperature measured at the fixture. The event is terminated once the threshold, set by the volume requirements for each event, is met. The solenoid is then opened to let the water drain. The same procedure takes place for all water draw events. To measure each draw correctly and terminate the event properly, the water events are scheduled such that they do not overlap. Flow meters are also used as redundant measurements. Figure 15 and Figure 16 show the water measurement apparatus in the master bathroom and the kitchen sink, respectively.

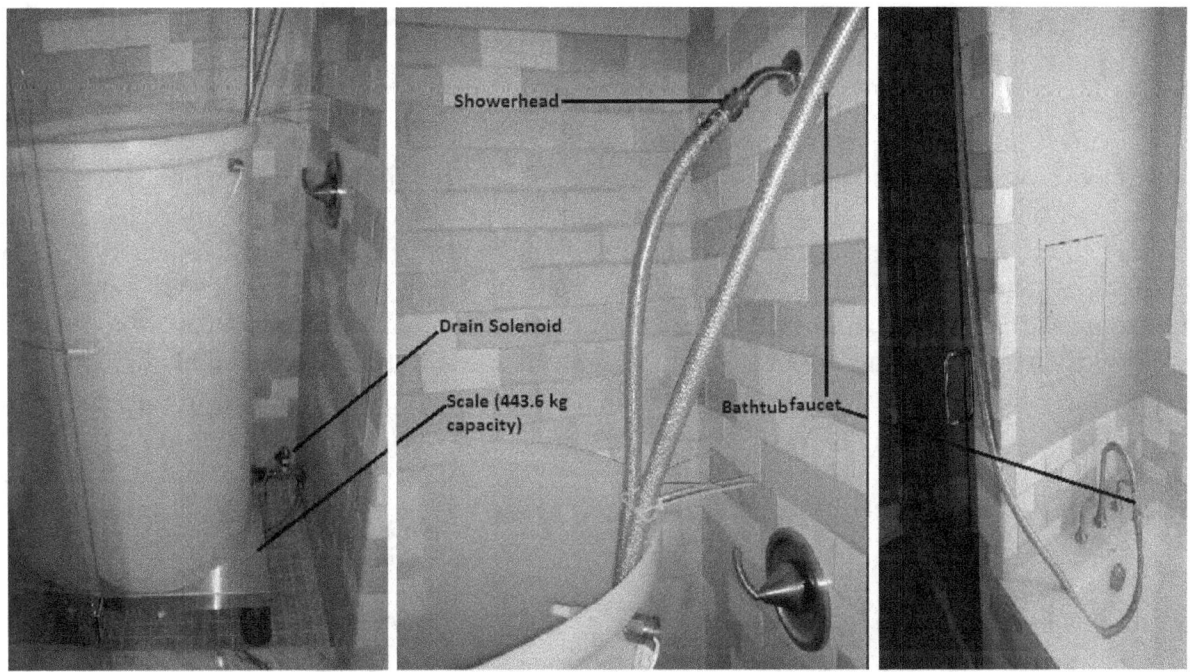

Figure 15. Master bathroom water measurement apparatus

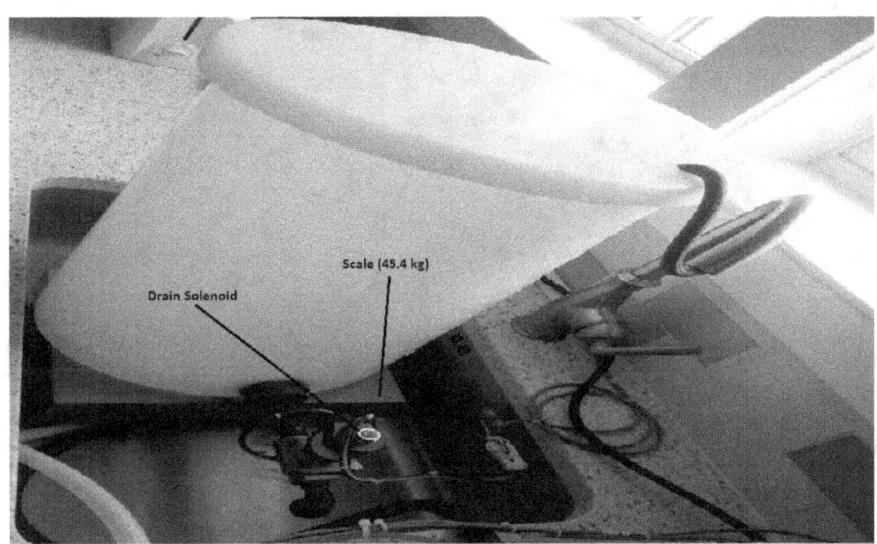

Figure 16. Kitchen sink water measurement apparatus

In order to determine the ratio of hot to cold water in each draw, the faucets were manually adjusted so that the water temperature exiting the fixtures met the ASHRAE's requirement for representative hot-water temperatures. According to Table 3 in the Service Water Heating chapter of the ASHRAE Handbook of HVAC applications, the hot water temperature for hand washing is 40.7 °C (105 °F), and 43.3 °C (110 °F) for showers and baths [12]. The heat pump water heater temperature is set to 48.9 °C (120 °F). These settings are adjusted throughout the year as the cold water temperature changes to ensure proper delivery temperature.

3.3 Miscellaneous Plug Loads

All plug loads are automated according to the schedule in Appendix A. The strategy to emulate plug loads (including cooking appliances) in the NZERTF is to use their annual energy consumption as a target and create a schedule to satisfy that requirement. The assumed frequency of usage of plug loads, in a given day, is based on the occupancy profile and may change from one day to the next. Some plug loads involve real appliances and others, which are difficult or unsafe to automate, are emulated with resistive loads. For example, modern televisions require activation with a remote control that is not easily automated, and a coffee maker is difficult to safely turn on and off using a data acquisition/controller unit. All plug loads, including standby, that are rated less than 200 W are emulated with resistant heating boxes similar to the one shown in Figure 13. However, plug loads larger than 200 W are emulated with heating elements connected to a relay-box shown in Figure 17. A relay-box, which energizes and de-energizes these plug loads and the real appliances, is shown in Figure 18.

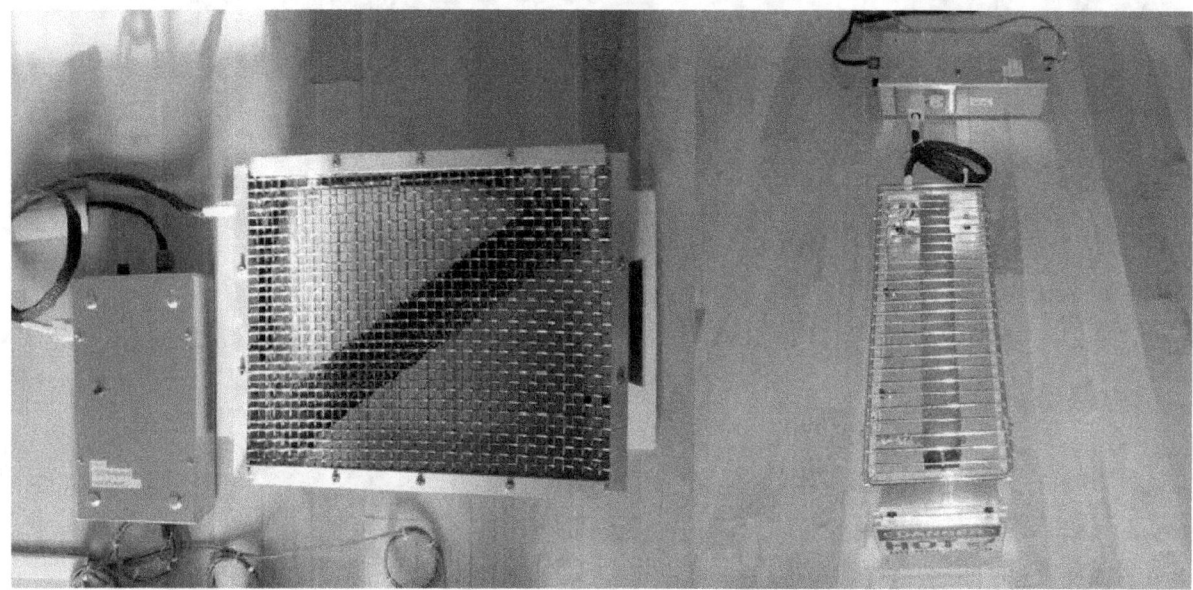

Figure 17. The heating elements for plug loads with larger than 200 W power requirement and their relay-boxes

20

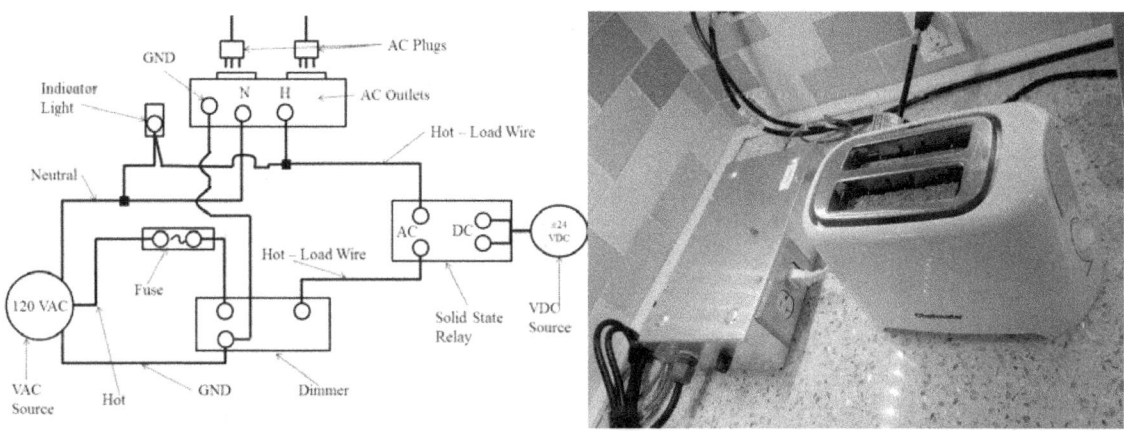

Figure 18. The relay-box, its wiring diagram, and a toaster plugged into it

The mechanisms to control all plug loads are separated into three main categories: time-based, criteria-based, and cycle-based. All plug loads, except the cycle-based loads, have a start-time and an end-time. In time-based control, the data acquisition/controller unit starts a load based on its start-time and turns it off when the end-time is reached. Similarly in criteria-based control the acquisition/controller unit starts a load based on its start-time, but it terminates the load when a certain criterion (e.g., energy consumption) has been met. For safety purposes, a timeout criterion is also applied to these loads which terminates their use if a certain elapsed time is reached. For example, when the hair dryer is energized, the data acquisition/controller unit monitors its energy consumption and terminates it when the energy consumed equals the prescribed energy. In criteria-based control, the end-time serves as a safety switch (a timeout) to prevent the loads from operating continuously in case of a malfunction beyond a predefined window. In the cycle-based control, the loads are activated based on their start-times and allowed to complete their normal cycles. The acquisition/controller unit does not terminate cycle-based loads. For implementation purposes, all plug loads are installed in various locations within the NZERTF. The location, control mechanism, emulation approach (actual appliance or heater box), active and standby power (where applicable), weekly time of use, and weekly required energy for each load are given in Table 10.

Table 10. The plug loads emulation details*

Rooms	Loads	Control Mechanism	Emulation Mechanism	Weekly Active Time of Use (h)	Estimated Standby Power (W)	Estimated Active Power (W)	Estimated Weekly Energy Consumption (Wh)
BR2[1]	Laptop	Time-based	HB[8]	25	3.1	36.9	1384.6
BR2	Boom Box	Time-based	HB	168	NA[10]	1.9	323.1
BR2	Cellphone Charger	Time-based	HB	168	NA	8.9	1488.5
BR2	Clock Radio	Time-based	HB	168	NA	1.7	286.5
BR3[2]	Laptop	Time-based	HB	12.75	5.9	36.9	1384.6
BR4[3]	Desktop Monitor + PC	Time-based	HB	41	16.0	101.6	6192.3
BR4	DSL Modem	Time-based	HB	168	NA	2.0	338.5
BR4	OTHER	Time-based	HB	168	NA	1.1	180.8
BR4	Printer	Time-based	HB	168	NA	4.5	750.0
BR4	Wireless Router	Time-based	HB	168	NA	24.0	4032.7
KA[4]	Can Opener	Time-based	HB	0.82	NA	70.0	57.7
LR[5]	Blu-ray Player (DVD)	Time-based	HB	3.04	8.8	17.0	1500.0
LR	Cable Box	Time-based	HB	168	NA	17.5	2936.5
LR	Clock	Time-based	HB	168	NA	3.0	500.0
LR	Component Stereo	Time-based	HB	168	NA	17.5	2942.3
LR	Phone/Answering Machine	Time-based	HB	168	NA	6.5	1090.4
LR	Television	Time-based	HB	37.25	1.0	62.2	2447.7
LR	Video Game Console	Time-based	HB	3	4.3	27.0	788.9
MBR[6]	Blu-ray Player	Time-based	HB	3.04	8.8	17.0	1500.0
MBR	Heating Pad	Time-based	HB	1.75	NA	33.0	57.7
MBR	Portable Fan	Time-based	HB	11	NA	19.8	217.3
MBR	Television	Time-based	HB	6.75	1.0	45.36	429.6
MBR	Cable Box	Time-based	HB	168	NA	17.5	2936.5
MBR	Cellphone Charger	Time-based	HB	168	NA	8.9	1488.5
MBR	Cellphone Charger	Time-based	HB	168	NA	8.9	1488.5
MBR	Clock Radio	Time-based	HB	168	NA	1.7	286.5
KA	Microwave	Cycle-based	AA[9]	1.5	NA	1850.0	2598.1
BR4	Vacuum	Criteria-based	HB	1.95	NA	542.0	1057.7
KA	Blender	Criteria-based	AA	0.84	NA	161.0	134.6
KA	Coffee Maker	Criteria-based	HB	2.14	NA	550.0	1176.9
KA	Hand Mixer	Criteria-based	AA	0.34	NA	113.0	38.5
KA	Slow Cooker	Criteria-based	AA	18.09	NA	17.0	307.7
KA	Toaster	Criteria-based	AA	1.24	NA	712.0	882.7
KA	Toaster Oven	Criteria-based	AA	0.67	NA	923.0	638.5
MBA[7]	Hair Dryer + Curling Iron	Criteria-based	AA	0.54	NA	1500.0	809.6
MBR	Iron	Criteria-based	AA	0.84	NA	1200.0	1013.5

[1] BR2 – Second Bedroom,   [4] KA – Kitchen,   [7] MBA – Mater Bedroom Bathroom

[2] BR3 – Third Bedroom,   [5] LR – Living Room,   [8] HB – Heater-Box

[3] BR4 – Bedroom Four,   [6] MBR – Master Bedroom,   [9] AA – Actual Appliance

[10] NA – None Applicable – The active and standby power for these plug loads are combined and emulated together.

* The numbers in Table 10 are used for target set-points and do not reflect the measurement accuracy

For the majority of plug loads listed in Table 10, their active power is computed using their weekly energy divided by their weekly active time of use. The weekly active time of use is based on the occupancy profile of the family, and the weekly energy consumption is derived from their annual energy divided by 52 weeks. For ease of implementation the standby power for these loads are combined with their active power. Some plug loads run continuously and some are activated intermittently; however, the goal is to meet their weekly and, ultimately, their annual energy consumption. The plug loads' active power, given in Table 10, meets that goal. The active power, for the criteria-based and the cycle-based plug loads, was determined by measurement.

For the loads with both active and standby power, given in Table 10, the sum of the standby energy and active energy consumption equals their weekly energy consumption. The active power for the laptop is computed as the average of fully-on charged (29.5 W), and fully-on charging (44.3 W) modes obtained from the Standby Power Summary Table of Lawrence Berkeley National Laboratory [13]. The active power for the Desktop Monitor + PC is the sum of active powers for a desktop computer and a liquid-crystal display (LCD) monitor. According to [13], the active power for an LCD display in on mode is 27.6 W and for a desktop computer in on/idle mode is approximately 74 W. The active power for the Blu-ray player is obtained from [5], and active power for the Video Game console was obtained from [13]. The active power for the televisions in the master bedroom and the living room were obtained from the ENERGY STAR Television product list [14]. For these plug loads, the standby power is computed as the difference between the weekly energy and active energy consumption divided by the weekly standby time of use.

3.4 Cooking Loads

The control mechanism for cooking events is similar to the criteria-based control approach described in the Section 3.3. Table 11 shows the list of cooking appliances that are installed in the NZERTF. In addition to the appliances, a range hood is also activated when the electric cooktop is activated. The range hood is set to its medium level and the active power consumption is, approximately, 100 W.

Table 11. Cooking appliances emulation details

Room	Loads	Control Mechanism	Emulation Mechanism	Weekly Frequency of Use	Estimated Weekly Energy Consumption (kWh)
Kitchen	Cooktop	Criteria-based	Actual Appliance	11	4.49
Kitchen	Oven	Criteria-based	Modified Appliance	3	5.84

* The estimated weekly energy in Table 11 are used for target set-points and does not reflect the measurement accuracy

The cooktop and the range hood are energized and de-energized using relay-boxes similar to the one shown in Figure 18. Figure 19 shows the cooktop installed in the NZERTF. Because of difficulties in controlling the built-in cooktop, a two-burner portable hot plate was used instead. Pots containing water are placed on top of the burners to absorb the heat that is produced. After they are energized, the energy consumption is monitored and the relay shuts off power when the prescribed amount of energy is used by the unit.

Figure 19. The electric two-burner portable hot plate, the range hood, and the relay-box

The NZERTF's oven has been modified to allow emulating cooking events. To safely operate the oven, a controller-box made of a proportional-integral-derivative controller, a limit controller switch, and relays was designed and built. The controller-box activates and de-activates the heating elements and a fan inside the oven to maintain a pre-set temperature of 176.6 °C (approximately 350 °F). The limit controller switch is installed to ensure that the oven does not exceed 250 °C (482 °F). The main software turns the controller-box off once the oven meets its pre-determined prescribed energy. Figure 20 shows the controller-box and its wiring schematic.

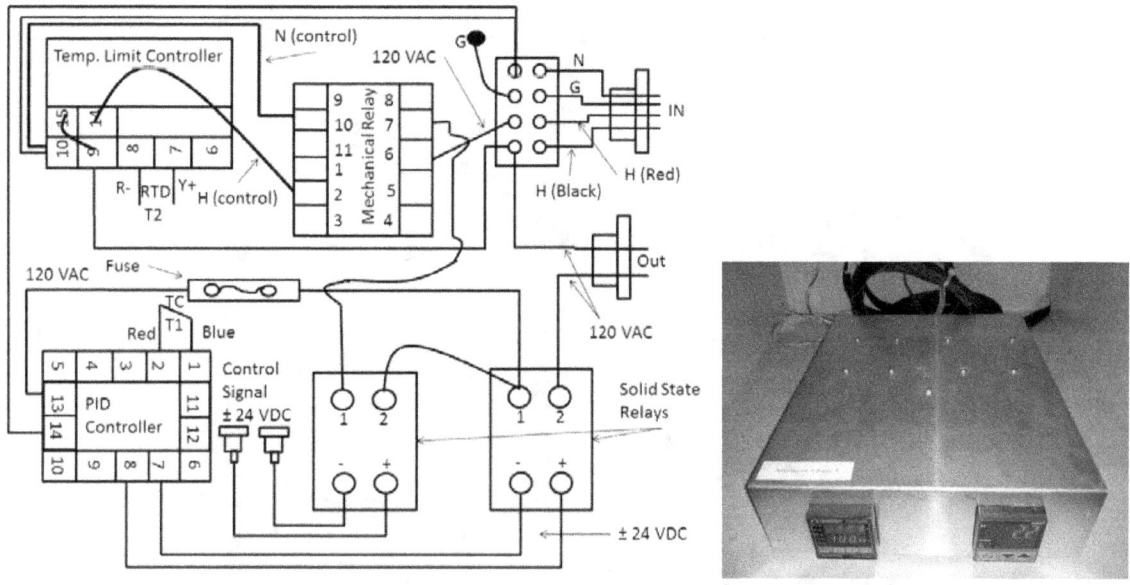

Figure 20. The oven Controller-box and its wiring schematic

3.5 Appliances

The clothes washer, clothes dryer, and dishwasher units were modified to allow remote activation. The clothes washer and dryer are directly controlled by the acquisition/controller unit with a dry-contact relay card, which initiates a normal cycle that automatically runs to completion. According to the Department of Energy (DOE) Table 5.1 of Section 2.8 of Appendix J2 to Subpart B of Part 430, Uniform Test Method for Measuring the Energy Consumption of Automatic and Semi-Automatic Clothes Washers, the average test load for a clothes washer container with 121.8 L (4.30 cu. ft.) capacity is 4.63 kg (10.20 lbs.) [15]. The clothes dryer is equipped with an advanced moisture sensing system. It is also equipped with a misting nozzle inside the drum container. Since placing a dry load of laundry ends the cycle prematurely, the misting nozzle was modified to allow water to be added to the laundry load. The volume of water added to the laundry was determined by washing the average test load and measuring the weight of the test load after the wash cycle. The final weight of the test load was measured to be 6.26 kg (13.80 lbs.). The misting nozzle flow rate was determined to be 3.79 liters (1 gallon) per minute. It is activated soon after the dryer cycle is energized and terminated when 1.63 L (1.63 kg) of water is dispensed inside the dryer. This procedure is applied to all clothes dryer cycles. Figure 21 shows the washer, dryer, and misting nozzle installed in the NZERTF.

Figure 21. The dryer, modified misting nozzle, and the washer

The dishwasher control is similar to the cycle-based control of plug loads described in Section 3.3.

The refrigerator installed in the NZERTF was tested, according to the Department of Energy (DOE) test procedure, to determine the value of the thermal load inside the refrigerator [16]. The annual energy consumption of this unit, according to the energy guide label, is 335 kWh/year and is determined by the DOE test procedure with no consumer interaction. To compensate for this, the DOE test uses elevated ambient temperature of 32.2 °C (90 °F). According to [16], approximately 5 kWh of the 335 kWh used per year is due to the periodic defrost cycles and the rest was used in the steady state operation of the unit. The study determined that running the thermal load (a heater) inside the refrigerator compartment at a constant (24 hours) value of 11.5 W brought the steady state back up to the annual energy consumption level (excluding the defrost). Therefore, we do not open and close the refrigerator door, except to remove ice, but instead we dissipate heat inside the refrigerator, to simulate the thermal load that people put in the refrigerator (e.g., groceries and warm food).

The energy use of the icemaker is not included in the annual energy consumption label. Operating the icemaker introduces a significant amount of moisture into the cabinet. The refrigerator has an adaptive defrost controller, so the added moisture from ice-making increases the number of the defrost cycles. It was decided that a researcher would enter the house and empty the ice bin once a week.

4. Conclusion

The NZERTF utilizes a virtual family to emulate water usage, lighting usage, miscellaneous electric plug loads, cooking loads, and appliance load. The virtual family, during the first year of testing, consists of two adults and two children whose behavior and activities are based on recommendations published by the U.S. Department of Energy. A minute-by-minute device usage schedule was developed to account for electrical energy consumption, sensible and latent heat generation, and hot water consumption. Some plug loads were emulated by resistive loads while others involve real appliances (e.g., microwave, toaster oven, etc.). Sensible loads were also emulated by resistive loads while latent load was emulated by an ultrasonic humidifier modified to enable long term operation without human intervention. The mechanisms to control sensible and latent, and electrical loads were separated into three categories: time-based, criteria-based, and cycle-based control. All loads, except the cycle-based loads, have a start-time and an end-time. In time-based control, the data acquisition/controller unit starts a load based on its start-time and turns it off when the end-time is reached. In criteria-based control the acquisition/controller unit starts a load based on its start-time, but it terminates the load when a certain criterion, e.g., energy consumption, has been met. In the cycle-based control, the loads were activated based on their start-times and allowed to complete their normal cycles. Water usage was emulated by energizing and de-energizing solenoid valves on the cold and hot water supply lines. The control mechanism is similar to the criteria-based control. The NZERTF is currently being used to demonstrate the feasibility of achieving net zero energy over the course a year (e.g., energy generated using photovoltaic modules and solar hot water panels equals energy consumed). It will subsequently serve as a test bed to develop measurement science, test/performance metric, for energy efficient building technologies. The load automation techniques discussed in the report can be utilized to emulate a range of occupant behavior.

References

[1]   J. Seryak and K. Kissock, "Occupancy and Behavioral Affects on Rresidential Enery Use," *Proceedings of annual conference on American solar energy society*, 2000.

[2]   R. Hendrom, "Building America Research Benchmark Definition Updated December 19 , 2008," National Renewable Energy Laboratory, 2008.

[3]   J. D. Lutz, A. Lekov, Y. Qin, and M. Melody, "Hot Water Draw Patterns in Single-Family Houses : Findings from Field Studies," Lawrence Berkeley National Laboratory, 2011.

[4]   R. Hendron and J. Burch, "Development of Standardized Domestic Hot Water Event Schedules for Residential Buildings," National Renewable Energy Laboratory, 2008.

[5]   L. International, D&R, *2009 Building Energy Data Book*. U.S. Depeartment of Energy, 2009.

[6]   T. and M. A. Group, "Technical Support Document: Energy Conservation Standards for Consumer Pproducts," 1998.

[7]   C. Iglehart, N. Milesi_Ferretti, and M. Galler, "Consumer Use of Dishwashers , Clothes Washers , and Dryers : Data Needs and Availability," NIST Technical Note 1696, 2011.

[8]   ASHRAE, *2013 ASHRAE Handbook - Fundamentals*. 2013.

[9]   J. L. Monteith, "Latent Heat of Vaporization in Thermal Physiology.," *Nature: New biology*, vol. 236, no. 64. p. 96, 22-Mar-1972.

[10]   R. Barnhart, "Home Moisture," Minnesota Department of Commerce, Last Modified August 2nd, 2012.

[11]   X. Fang, D. Christensen, G. Barker, and E. Hancock, "Field Test Protocol : Standard Internal Load Generation for Unoccupied Test Homes," Building Technologies Program, U.S. Department of ENERGY, 2011.

[12]   ASHRAE, *2011 ASHRAE Handbook - HVAC Applications*. 2011.

[13]   LBNL, "Standby Power Summary Table." Lawrence Berkeley National Laboratory, http://standby.lbl.gov/summary-table.html.

[14]   "ENERGY STAR Television Product List." ENERGY STAR, http://downloads.energystar.gov/bi/qplist/tv_prod_list.pdf, 2013.

[15]   "Uniform Test Method for Measuring the Energy Consumption of Automatic and Semi-Automatic Clothes Washers." 10 CFR 430. APPENDIX J2. SUBPART B, 2013.

[16]   D. Yashar, "Refrigerator's Thermal Load," Private Communications, National Institute of Standards and Technology, MD, 2012.

Appendix A

For the loads that are time-based controlled (see Section 3.3 for details), the start-time and end-time are given but the corresponding entries in columns labeled as (Target Volume / Draw) and (Target Energy / Event) are left blank. For the loads that are criteria-based controlled (e.g., a sink use or a blender use) start-time, end-time, target volume, and target energy are given in columns labeled Target Volume / Draw and Target Energy / Event. The end-time is a safety timeout mechanism. For the loads that are cycle-based controlled only the start-time has been given, and all other corresponding entries are left blank. The numbers (e.g., target volume and energy) throughout the schedule are calculations used for target set-points and do not reflect the measurement accuracy.

Mondays

Start time	End time	FixtureID	Description	Target Volume / Draw (liters)	Target Volume / Draw (gallons)	Target Energy / Day (Wh)
12:05:00 AM	6:00:00 AM	SHparentBMBR	Sensible Heat ParentB Master Bedroom			
12:05:00 AM	6:30:00 AM	SHparentAMBR	Sensible Heat ParentA Master Bedroom			
12:05:00 AM	7:00:00 AM	SHchildABR2	Sensible Heat ChildA Bedroom # 2			
12:05:00 AM	7:00:00 AM	SHchildBBR3	Sensible Heat ChildB Bedroom # 3			
12:05:00 AM	8:30:00 AM	Latent Load	Latent Load			
3:31:00 AM	3:34:00 AM	LightMastBath	Master Bathroom Lights			
3:33:00 AM	3:34:00 AM	SinkMastBathH	Sink event - Master Bathroom hot water	2.38	0.63	
4:00:00 AM	4:03:00 AM	LightMastBath	Master Bathroom Lights			
4:02:00 AM	4:03:00 AM	SinkMastBathH	Sink event - Master Bathroom hot water	2.38	0.63	
6:00:00 AM	6:01:00 AM	KitchenSinkH	Kitchen Sink hot water	2.38	0.63	
6:00:00 AM	6:15:00 AM	LightMastBath	Master Bathroom Lights			
6:00:00 AM	6:30:00 AM	SHparentBDR	Sensible Heat ParentB Dining Room			
6:00:00 AM	8:25:00 AM	KitchenLights	Kitchen Lights			
6:00:00 AM		MicrowaveCLR	Clear Microwave (equivalent to opening and closing the door)			
6:01:00 AM		MicrowaveON	Turn Microwave ON			
6:04:00 AM	6:19:00 AM	ShowerMastBathH	Shower even - Master Bathroom hot water	53.00	14.00	
6:05:00 AM	6:15:00 AM	Blender	Blender			19.20
6:10:00 AM	6:28:20 AM	CoffeeMaker	Coffee Maker			168.12
6:15:00 AM	6:25:00 AM	ToasterOven	Toaster Oven			88.80
6:15:00 AM	6:30:00 AM	LightMastBed	Master Bedroom Lights			
6:15:00 AM	7:00:00 AM	MastBedATV	Master Bedroom Active TV			

6:20:00 AM	6:21:00 AM	SinkMastBathH	Sink event - Master Bathroom hot water	2.38	0.63	
6:20:00 AM	6:35:00 AM	Toaster	Toaster			126.00
6:25:00 AM	6:26:00 AM	KitchenSinkH	Kitchen Sink hot water	2.38	0.63	
6:30:00 AM	6:45:00 AM	LightMastBath	Master Bathroom Lights			
6:30:00 AM	6:50:00 AM	Light2ndBath	2nd Bathroom Lights			
6:30:00 AM	7:00:00 AM	SHparentBMBR	Sensible Heat ParentB Master Bedroom			
6:30:00 AM	8:25:00 AM	LivingRATV	Living room - Active TV			
6:30:00 AM	8:30:00 AM	SHparentAK	Sensible Heat ParentA Kitchen			
6:32:00 AM	6:42:00 AM	ShowerMastBathH	Shower even - Master Bathroom hot water	33.12	8.75	
6:40:00 AM	6:50:00 AM	HairDryer	Hair Dryer			115.70
6:40:00 AM	7:00:00 AM	Light2ndBed	2nd Bedroom Lights			
6:43:00 AM	6:44:00 AM	SinkMastBathH	Sink event - Master Bathroom hot water	2.38	0.63	
6:45:00 AM	6:55:00 AM	ShowerMastBathH	Shower even - Master Bathroom hot water	33.12	8.75	
6:45:00 AM	7:00:00 AM	LightMastBed	Master Bedroom Lights			
6:50:00 AM	7:00:00 AM	Light3rdBed	3rd Bedroom Lights			
6:56:00 AM	6:57:00 AM	SinkMastBathH	Sink event - Master Bathroom hot water	2.38	0.63	
6:58:00 AM	6:59:00 AM	SinkMastBathH	Sink event - Master Bathroom hot water	2.38	0.63	
7:00:00 AM	8:00:00 AM	DiningRLights	Dining Room Lights			
7:00:00 AM	8:00:00 AM	SHchildALR	Sensible Heat ChildA Living Room			
7:00:00 AM	8:00:00 AM	SHchildBDR	Sensible Heat ChildB Dining Room			
7:00:00 AM	8:30:00 AM	SHparentBDR	Sensible Heat ParentB Dining Room			
7:05:00 AM	7:06:00 AM	KitchenSinkH	Kitchen Sink hot water	2.38	0.63	
7:30:00 AM	7:31:00 AM	KitchenSinkH	Kitchen Sink hot water	2.38	0.63	
8:00:00 AM	8:25:00 AM	LivingRLights	Living room Lights			
8:02:00 AM	8:03:00 AM	KitchenSinkH	Kitchen Sink hot water	2.38	0.63	
8:10:00 AM	8:11:00 AM	KitchenSinkH	Kitchen Sink hot water	2.38	0.63	
8:15:00 AM	8:20:00 AM	Light1stBath	1st Bathroom Lights			
8:20:00 AM	8:21:00 AM	KitchenSinkH	Kitchen Sink hot water	2.38	0.63	
8:26:00 AM	8:27:00 AM	KitchenSinkH	Kitchen Sink hot water	2.38	0.63	
8:28:00 AM	8:29:00 AM	KitchenSinkH	Kitchen Sink hot water	2.38	0.63	
4:00:00 PM	11:59:00 PM	Latent Load	Latent Load			
6:00:00 PM	6:30:00 PM	KitchenLights	Kitchen Lights			
6:00:00 PM	6:30:00 PM	SHchildBDR	Sensible Heat ChildB Dining Room			
6:00:00 PM	6:30:00 PM	SHparentAMBR	Sensible Heat ParentA Master Bedroom			
6:00:00 PM	7:30:00 PM	SHchildABR2	Sensible Heat ChildA Bedroom # 2			
6:00:00 PM	10:00:00 PM	SHparentBDR	Sensible Heat ParentB Dining Room			
6:00:00 PM	10:30:00 PM	LivingRLights	Living room Lights			
6:01:00 PM	6:06:00 PM	LightMastBath	Master Bathroom Lights			

6:02:00 PM	6:06:00 PM	Light1stBath	1st Bathroom Lights			
6:02:00 PM	7:02:00 PM	CookTops	Cooktops			408.04
6:03:00 PM	6:04:00 PM	KitchenSinkH	Kitchen Sink hot water	2.38	0.63	
6:05:00 PM	6:06:00 PM	SinkMastBathH	Sink event - Master Bathroom hot water	2.38	0.63	
6:07:00 PM	6:08:00 PM	KitchenSinkH	Kitchen Sink hot water	2.38	0.63	
6:07:00 PM	6:30:00 PM	LightMastBed	Master Bedroom Lights			
6:08:00 PM	6:18:00 PM	Light2ndBath	2nd Bathroom Lights			
6:09:00 PM	6:19:00 PM	ShowerMastBathH	Shower even - Master Bathroom hot water	33.12	8.75	
6:09:00 PM		MicrowaveCLR	Clear Microwave (equivalent to opening and closing the door)			
6:10:00 PM	8:30:00 PM	LivingRATV	Living room - Active TV			
6:10:00 PM		MicrowaveON	Turn Microwave ON			
6:20:00 PM	6:21:00 PM	SinkMastBathH	Sink event - Master Bathroom hot water	2.38	0.63	
6:20:00 PM	7:30:00 PM	Light2ndBed	2nd Bedroom Lights			
6:30:00 PM	7:30:00 PM	Light3rdBed	3rd Bedroom Lights			
6:30:00 PM	7:30:00 PM	SHchildBBR3	Sensible Heat ChildB Bedroom # 3			
6:30:00 PM	8:15:00 PM	ComputerACB	Bedroom # 3 - Laptop			
6:30:00 PM	10:00:00 PM	ComputerACA	Bedroom # 2 - Laptop			
6:30:00 PM	10:30:00 PM	DesktopCA	Desktop PC and Monitor			342.60
6:30:00 PM	10:30:00 PM	SHparentAK	Sensible Heat ParentA Kitchen			
6:40:00 PM	6:41:00 PM	KitchenSinkH	Kitchen Sink hot water	2.38	0.63	
6:40:00 PM	6:47:04 PM	CanOpener	Can Opener			
6:40:00 PM	6:50:00 PM	HandMixer	Hand Mixer			5.50
6:45:00 PM	6:46:00 PM	KitchenSinkH	Kitchen Sink hot water	2.38	0.63	
6:55:00 PM	6:56:00 PM	KitchenSinkH	Kitchen Sink hot water	2.38	0.63	
7:00:00 PM	7:15:00 PM	KitchenLights	Kitchen Lights			
7:05:00 PM	7:06:00 PM	KitchenSinkH	Kitchen Sink hot water	2.38	0.63	
7:10:00 PM	7:11:00 PM	KitchenSinkH	Kitchen Sink hot water	2.38	0.63	
7:15:00 PM	8:30:00 PM	DiningRLights	Dining Room Lights			
7:30:00 PM	8:15:00 PM	KitchenLights	Kitchen Lights			
7:30:00 PM	8:15:00 PM	SHchildALR	Sensible Heat ChildA Living Room			
7:30:00 PM	8:15:00 PM	SHchildBDR	Sensible Heat ChildB Dining Room			
7:35:00 PM	7:36:00 PM	KitchenSinkH	Kitchen Sink hot water	2.38	0.63	
8:00:00 PM	8:01:00 PM	KitchenSinkH	Kitchen Sink hot water	2.38	0.63	
8:15:00 PM	8:16:00 PM	KitchenSinkH	Kitchen Sink hot water	2.38	0.63	
8:15:00 PM	8:25:00 PM	Light2ndBath	2nd Bathroom Lights			
8:15:00 PM	8:30:00 PM	Light3rdBed	3rd Bedroom Lights			
8:15:00 PM	9:45:00 PM	Light2ndBed	2nd Bedroom Lights			
8:15:00 PM	11:59:00 PM	SHchildABR2	Sensible Heat ChildA Bedroom # 2			
8:15:00 PM	11:59:00 PM	SHchildBBR3	Sensible Heat ChildB Bedroom # 3			

8:20:00 PM	8:21:00 PM	SinkMastBathH	Sink event - Master Bathroom hot water	2.38	0.63	
8:25:00 PM	8:26:00 PM	KitchenSinkH	Kitchen Sink hot water	2.38	0.63	
8:28:00 PM		Dishwasher	Dishwasher			
8:30:00 PM	10:30:00 PM	KitchenLights	Kitchen Lights			
8:35:00 PM	8:36:00 PM	KitchenSinkH	Kitchen Sink hot water	2.38	0.63	
8:50:00 PM	8:51:00 PM	KitchenSinkH	Kitchen Sink hot water	2.38	0.63	
9:05:00 PM	9:06:00 PM	KitchenSinkH	Kitchen Sink hot water	2.38	0.63	
9:10:00 PM	9:11:00 PM	KitchenSinkH	Kitchen Sink hot water	2.38	0.63	
9:30:00 PM	9:31:00 PM	KitchenSinkH	Kitchen Sink hot water	2.38	0.63	
9:35:00 PM	9:36:00 PM	KitchenSinkH	Kitchen Sink hot water	2.38	0.63	
9:45:00 PM	10:00:00 PM	Light2ndBath	2nd Bathroom Lights			
9:55:00 PM	9:56:00 PM	SinkMastBathH	Sink event - Master Bathroom hot water	2.38	0.63	
10:00:00 PM	10:15:00 PM	HeatingPad	Heating Pad			
10:00:00 PM	11:00:00 PM	LightMastBed	Master Bedroom Lights			
10:00:00 PM	11:59:00 PM	PortableFan	Portable Fan			
10:00:00 PM	11:59:00 PM	SHparentBMBR	Sensible Heat ParentB Master Bedroom			
10:05:00 PM	10:06:00 PM	KitchenSinkH	Kitchen Sink hot water	2.38	0.63	
10:05:00 PM	10:25:00 PM	LightMastBath	Master Bathroom Lights			
10:10:00 PM	10:11:00 PM	KitchenSinkH	Kitchen Sink hot water	2.38	0.63	
10:20:00 PM	10:21:00 PM	SinkMastBathH	Sink event - Master Bathroom hot water	2.38	0.63	
10:30:00 PM	10:40:00 PM	LightMastBath	Master Bathroom Lights			
10:30:00 PM	11:59:00 PM	SHparentAMBR	Sensible Heat ParentA Master Bedroom			
10:35:00 PM	10:36:00 PM	SinkMastBathH	Sink event - Master Bathroom hot water	2.38	0.63	

Tuesdays

Start time	End time	FixtureID	Description	Target Volume / Draw (liters)	Target Volume / Draw (gallons)	Target Energy / Event (Wh)
12:05:00 AM	6:00:00 AM	SHparentBMBR	Sensible Heat ParentB Master Bedroom			
12:05:00 AM	6:30:00 AM	SHparentAMBR	Sensible Heat ParentA Master Bedroom			
12:05:00 AM	7:00:00 AM	SHchildABR2	Sensible Heat ChildA Bedroom # 2			
12:05:00 AM	7:00:00 AM	SHchildBBR3	Sensible Heat ChildB Bedroom # 3			
12:05:00 AM	8:30:00 AM	Latent Load	Latent Load			
3:31:00 AM	3:34:00 AM	LightMastBath	Master Bathroom Lights			
3:33:00 AM	3:34:00 AM	SinkMastBathH	Sink event - Master Bathroom hot water	2.38	0.63	
4:00:00 AM	4:03:00 AM	LightMastBath	Master Bathroom Lights			
4:02:00 AM	4:03:00 AM	SinkMastBathH	Sink event - Master Bathroom hot water	2.38	0.63	
6:00:00 AM	6:01:00 AM	KitchenSinkH	Kitchen Sink hot water	2.38	0.63	
6:00:00 AM	6:15:00 AM	LightMastBath	Master Bathroom Lights			
6:00:00 AM	6:30:00 AM	SHparentBDR	Sensible Heat ParentB Dining Room			
6:00:00 AM	8:25:00 AM	KitchenLights	Kitchen Lights			
6:00:00 AM		MicrowaveCLR	Clear Microwave (equivalent to opening and closing the door)			
6:01:00 AM		MicrowaveON	Turn Microwave ON			
6:02:00 AM	6:03:00 AM	KitchenSinkH	Kitchen Sink hot water	2.38	0.63	
6:04:00 AM	6:14:00 AM	ShowerMastBathH	Shower even - Master Bathroom hot water	33.12	8.75	
6:05:00 AM	6:15:00 AM	Blender	Blender			19.20
6:10:00 AM	6:40:00 AM	CoffeeMaker	Coffee Maker			168.12
6:15:00 AM	6:16:00 AM	SinkMastBathH	Sink event - Master Bathroom hot water	2.38	0.63	
6:15:00 AM	6:25:00 AM	ToasterOven	Toaster Oven			88.80
6:15:00 AM	6:30:00 AM	LightMastBed	Master Bedroom Lights			
6:15:00 AM	7:00:00 AM	MastBedATV	Master Bedroom Active TV			
6:20:00 AM	6:35:00 AM	Toaster	Toaster			126.00
6:25:00 AM	6:26:00 AM	KitchenSinkH	Kitchen Sink hot water	2.38	0.63	
6:30:00 AM	6:40:00 AM	ShowerMastBathH	Shower even - Master Bathroom hot water	33.12	8.75	
6:30:00 AM	6:45:00 AM	LightMastBath	Master Bathroom Lights			
6:30:00 AM	6:50:00 AM	Light2ndBath	2nd Bathroom Lights			
6:30:00 AM	7:00:00 AM	SHparentBMBR	Sensible Heat ParentB Master Bedroom			
6:30:00 AM	8:25:00 AM	LivingRATV	Living room - Active TV			
6:30:00 AM	8:30:00 AM	SHparentAK	Sensible Heat ParentA Kitchen			
6:40:00 AM	6:50:00 AM	HairDryer	Hair Dryer			115.70
6:40:00 AM	7:00:00 AM	Light2ndBed	2nd Bedroom Lights			

6:41:00 AM	6:42:00 AM	SinkMastBathH	Sink event - Master Bathroom hot water	2.38	0.63	
6:43:00 AM	6:44:00 AM	KitchenSinkH	Kitchen Sink hot water	2.38	0.63	
6:45:00 AM	6:55:00 AM	ShowerMastBathH	Shower even - Master Bathroom hot water	33.12	8.75	
6:45:00 AM	7:00:00 AM	LightMastBed	Master Bedroom Lights			
6:50:00 AM	7:00:00 AM	Light3rdBed	3rd Bedroom Lights			
6:56:00 AM	6:57:00 AM	SinkMastBathH	Sink event - Master Bathroom hot water	2.38	0.63	
6:58:00 AM	6:59:00 AM	SinkMastBathH	Sink event - Master Bathroom hot water	2.38	0.63	
7:00:00 AM	8:00:00 AM	DiningRLights	Dining Room Lights			
7:00:00 AM	8:00:00 AM	SHchildALR	Sensible Heat ChildA Living Room			
7:00:00 AM	8:00:00 AM	SHchildBDR	Sensible Heat ChildB Dining Room			
7:00:00 AM	8:30:00 AM	SHparentBDR	Sensible Heat ParentB Dining Room			
7:05:00 AM	7:06:00 AM	KitchenSinkH	Kitchen Sink hot water	2.38	0.63	
7:30:00 AM	7:31:00 AM	KitchenSinkH	Kitchen Sink hot water	2.38	0.63	
7:35:00 AM	7:36:00 AM	KitchenSinkH	Kitchen Sink hot water	2.38	0.63	
8:00:00 AM	8:25:00 AM	LivingRLights	Living room Lights			
8:02:00 AM	8:03:00 AM	KitchenSinkH	Kitchen Sink hot water	2.38	0.63	
8:10:00 AM	8:11:00 AM	KitchenSinkH	Kitchen Sink hot water	2.38	0.63	
8:20:00 AM	8:21:00 AM	KitchenSinkH	Kitchen Sink hot water	2.38	0.63	
8:26:00 AM	8:27:00 AM	KitchenSinkH	Kitchen Sink hot water	2.38	0.63	
8:30:00 AM	8:30:00 PM	SlowCooker	Slow Cooker			153.85
4:00:00 PM	4:30:00 PM	SHchildALR	Sensible Heat ChildA Living Room			
4:00:00 PM	11:59:00 PM	Latent Load	Latent Load			
4:01:00 PM	4:06:00 PM	Light1stBath	1st Bathroom Lights			
4:04:00 PM	4:05:00 PM	KitchenSinkH	Kitchen Sink hot water	2.38	0.63	
4:10:00 PM	4:11:00 PM	KitchenSinkH	Kitchen Sink hot water	2.38	0.63	
4:25:00 PM	4:26:00 PM	KitchenSinkH	Kitchen Sink hot water	2.38	0.63	
4:30:00 PM	6:00:00 PM	Light2ndBed	2nd Bedroom Lights			
4:30:00 PM	6:00:00 PM	SHchildABR2	Sensible Heat ChildA Bedroom # 2			
6:00:00 PM	6:30:00 PM	KitchenLights	Kitchen Lights			
6:00:00 PM	6:30:00 PM	LivingRLights	Living room Lights			
6:00:00 PM	6:30:00 PM	SHchildALR	Sensible Heat ChildA Living Room			
6:00:00 PM	6:30:00 PM	SHchildBDR	Sensible Heat ChildB Dining Room			
6:00:00 PM	6:30:00 PM	SHparentAMBR	Sensible Heat ParentA Master Bedroom			
6:00:00 PM	10:00:00 PM	SHparentBDR	Sensible Heat ParentB Dining Room			
6:01:00 PM	6:06:00 PM	LightMastBath	Master Bathroom Lights			
6:02:00 PM	6:06:00 PM	Light1stBath	1st Bathroom Lights			
6:02:00 PM	7:02:00 PM	CookTops	Cooktops			408.04

6:05:00 PM	6:06:00 PM	SinkMastBathH	Sink event - Master Bathroom hot water	2.38	0.63	
6:07:00 PM	6:30:00 PM	LightMastBed	Master Bedroom Lights			
6:09:00 PM		MicrowaveCLR	Clear Microwave (equivalent to opening and closing the door)			
6:10:00 PM	8:30:00 PM	LivingRATV	Living room - Active TV			
6:10:00 PM		MicrowaveON	Turn Microwave ON			
6:30:00 PM	7:30:00 PM	SHchildABR2	Sensible Heat ChildA Bedroom # 2			
6:30:00 PM	7:30:00 PM	SHchildBBR3	Sensible Heat ChildB Bedroom # 3			
6:30:00 PM	8:15:00 PM	ComputerACB	Bedroom # 3 - Laptop			
6:30:00 PM	8:30:00 PM	Light3rdBed	3rd Bedroom Lights			
6:30:00 PM	10:00:00 PM	ComputerACA	Bedroom # 2 - Laptop			
6:30:00 PM	10:00:00 PM	Light2ndBed	2nd Bedroom Lights			
6:30:00 PM	10:30:00 PM	DesktopCA	Desktop PC and Monitor			342.60
6:30:00 PM	10:30:00 PM	SHparentAK	Sensible Heat ParentA Kitchen			
6:40:00 PM	6:41:00 PM	KitchenSinkH	Kitchen Sink hot water	2.38	0.63	
6:40:00 PM	6:47:04 PM	CanOpener	Can Opener			
6:40:00 PM	6:50:00 PM	HandMixer	Hand Mixer			5.50
6:45:00 PM	6:46:00 PM	KitchenSinkH	Kitchen Sink hot water	2.38	0.63	
6:55:00 PM	6:56:00 PM	KitchenSinkH	Kitchen Sink hot water	2.38	0.63	
7:00:00 PM	7:15:00 PM	KitchenLights	Kitchen Lights			
7:00:00 PM	7:15:00 PM	LivingRLights	Living room Lights			
7:10:00 PM	7:11:00 PM	KitchenSinkH	Kitchen Sink hot water	2.38	0.63	
7:15:00 PM	8:30:00 PM	DiningRLights	Dining Room Lights			
7:30:00 PM	8:15:00 PM	KitchenLights	Kitchen Lights			
7:30:00 PM	8:15:00 PM	SHchildALR	Sensible Heat ChildA Living Room			
7:30:00 PM	8:15:00 PM	SHchildBDR	Sensible Heat ChildB Dining Room			
7:30:00 PM	9:30:00 PM	LivingRLights	Living room Lights			
7:35:00 PM	7:36:00 PM	KitchenSinkH	Kitchen Sink hot water	2.38	0.63	
8:00:00 PM	8:01:00 PM	KitchenSinkH	Kitchen Sink hot water	2.38	0.63	
8:15:00 PM	8:16:00 PM	KitchenSinkH	Kitchen Sink hot water	2.38	0.63	
8:15:00 PM	8:25:00 PM	Light2ndBath	2nd Bathroom Lights			
8:15:00 PM	11:59:00 PM	SHchildABR2	Sensible Heat ChildA Bedroom # 2			
8:15:00 PM	11:59:00 PM	SHchildBBR3	Sensible Heat ChildB Bedroom # 3			
8:20:00 PM	8:21:00 PM	SinkMastBathH	Sink event - Master Bathroom hot water	2.38	0.63	
8:25:00 PM	8:26:00 PM	KitchenSinkH	Kitchen Sink hot water	2.38	0.63	
8:30:00 PM	10:30:00 PM	KitchenLights	Kitchen Lights			
8:35:00 PM	8:36:00 PM	KitchenSinkH	Kitchen Sink hot water	2.38	0.63	
8:50:00 PM	8:51:00 PM	KitchenSinkH	Kitchen Sink hot water	2.38	0.63	
9:05:00 PM	9:06:00 PM	KitchenSinkH	Kitchen Sink hot water	2.38	0.63	
9:10:00 PM	9:11:00 PM	KitchenSinkH	Kitchen Sink hot water	2.38	0.63	

9:30:00 PM	9:31:00 PM	KitchenSinkH	Kitchen Sink hot water	2.38	0.63	
9:40:00 PM	9:41:00 PM	KitchenSinkH	Kitchen Sink hot water	2.38	0.63	
9:43:00 PM	9:58:00 PM	ShowerMastBathH	Shower even - Master Bathroom hot water	53.00	14.00	
9:45:00 PM	10:00:00 PM	Light2ndBath	2nd Bathroom Lights			
9:59:00 PM	10:00:00 PM	SinkMastBathH	Sink event - Master Bathroom hot water	2.38	0.63	
10:00:00 PM	12:00:00 AM	PortableFan	Portable Fan			
10:00:00 PM	10:10:00 PM	HeatingPad	Heating Pad			
10:00:00 PM	10:30:00 PM	LivingRLights	Living room Lights			
10:00:00 PM	11:00:00 PM	LightMastBed	Master Bedroom Lights			
10:00:00 PM	11:59:00 PM	SHparentBMBR	Sensible Heat ParentB Master Bedroom			
10:05:00 PM	10:06:00 PM	KitchenSinkH	Kitchen Sink hot water	2.38	0.63	
10:05:00 PM	10:25:00 PM	LightMastBath	Master Bathroom Lights			
10:20:00 PM	10:21:00 PM	SinkMastBathH	Sink event - Master Bathroom hot water	2.38	0.63	
10:30:00 PM	10:40:00 PM	LightMastBath	Master Bathroom Lights			
10:30:00 PM	11:59:00 PM	SHparentAMBR	Sensible Heat ParentA Master Bedroom			
10:35:00 PM	10:36:00 PM	SinkMastBathH	Sink event - Master Bathroom hot water	2.38	0.63	

Wednesdays

Start time	End time	FixtureID	Description	Target volume / Draw (liters)	Target Volume / Draw (gallons)	Target Energy / Event (Wh)
12:05:00 AM	6:00:00 AM	SHparentBMBR	Sensible Heat ParentB Master Bedroom			
12:05:00 AM	6:30:00 AM	SHparentAMBR	Sensible Heat ParentA Master Bedroom			
12:05:00 AM	7:00:00 AM	SHchildABR2	Sensible Heat ChildA Bedroom # 2			
12:05:00 AM	7:00:00 AM	SHchildBBR3	Sensible Heat ChildB Bedroom # 3			
12:05:00 AM	8:30:00 AM	Latent Load	Latent Load			
3:31:00 AM	3:34:00 AM	LightMastBath	Master Bathroom Lights			
3:33:00 AM	3:34:00 AM	SinkMastBathH	Sink event - Master Bathroom hot water	2.38	0.63	
4:00:00 AM	4:03:00 AM	LightMastBath	Master Bathroom Lights			
4:02:00 AM	4:03:00 AM	SinkMastBathH	Sink event - Master Bathroom hot water	2.38	0.63	
6:00:00 AM	6:01:00 AM	KitchenSinkH	Kitchen Sink hot water	2.38	0.63	
6:00:00 AM	6:15:00 AM	LightMastBath	Master Bathroom Lights			
6:00:00 AM	6:30:00 AM	SHparentBDR	Sensible Heat ParentB Dining Room			
6:00:00 AM	8:25:00 AM	KitchenLights	Kitchen Lights			
6:00:00 AM		MicrowaveCLR	Clear Microwave (equivalent to opening and closing the door)			
6:01:00 AM		MicrowaveON	Turn Microwave ON			
6:03:00 AM	6:13:00 AM	ShowerMastBathH	Shower even - Master Bathroom hot water	33.12	8.75	
6:05:00 AM	6:15:00 AM	Blender	Blender			19.20
6:09:00 AM		MicrowaveCLR	Clear Microwave (equivalent to opening and closing the door)			
6:10:00 AM	6:40:00 AM	CoffeeMaker	Coffee Maker			168.12
6:14:00 AM	6:15:00 AM	SinkMastBathH	Sink event - Master Bathroom hot water	2.38	0.63	
6:15:00 AM	6:25:00 AM	ToasterOven	Toaster Oven			88.80
6:15:00 AM	6:30:00 AM	LightMastBed	Master Bedroom Lights			
6:15:00 AM	7:00:00 AM	MastBedATV	Master Bedroom Active TV			
6:20:00 AM	6:35:00 AM	Toaster	Toaster			126.00
6:25:00 AM	6:26:00 AM	KitchenSinkH	Kitchen Sink hot water	2.38	0.63	
6:30:00 AM	6:40:00 AM	Light2ndBath	2nd Bathroom Lights			
6:30:00 AM	6:45:00 AM	LightMastBath	Master Bathroom Lights			
6:30:00 AM	7:00:00 AM	SHparentBMBR	Sensible Heat ParentB Master Bedroom			
6:30:00 AM	8:25:00 AM	LivingRATV	Living room - Active TV			
6:30:00 AM	8:30:00 AM	SHparentAK	Sensible Heat ParentA Kitchen			
6:31:00 AM	6:41:00 AM	ShowerMastBathH	Shower even - Master Bathroom hot water	33.12	8.75	

6:40:00 AM	6:50:00 AM	HairDryer	Hair Dryer			115.70
6:40:00 AM	7:00:00 AM	Light2ndBed	2nd Bedroom Lights			
6:42:00 AM	6:43:00 AM	SinkMastBathH	Sink event - Master Bathroom hot water	2.38	0.63	
6:44:00 AM	6:45:00 AM	SinkMastBathH	Sink event - Master Bathroom hot water	2.38	0.63	
6:45:00 AM	7:00:00 AM	LightMastBed	Master Bedroom Lights			
6:47:00 AM	6:48:00 AM	KitchenSinkH	Kitchen Sink hot water	2.38	0.63	
6:50:00 AM	7:00:00 AM	Light3rdBed	3rd Bedroom Lights			
7:00:00 AM	8:00:00 AM	DiningRLights	Dining Room Lights			
7:00:00 AM	8:00:00 AM	SHchildALR	Sensible Heat ChildA Living Room			
7:00:00 AM	8:00:00 AM	SHchildBDR	Sensible Heat ChildB Dining Room			
7:00:00 AM	8:30:00 AM	SHparentBDR	Sensible Heat ParentB Dining Room			
7:05:00 AM	7:06:00 AM	KitchenSinkH	Kitchen Sink hot water	2.38	0.63	
7:30:00 AM	7:31:00 AM	KitchenSinkH	Kitchen Sink hot water	2.38	0.63	
7:40:00 AM	7:45:00 AM	Light1stBath	1st Bathroom Lights			
7:42:00 AM	7:43:00 AM	KitchenSinkH	Kitchen Sink hot water	2.38	0.63	
7:45:00 AM	7:46:00 AM	KitchenSinkH	Kitchen Sink hot water	2.38	0.63	
8:00:00 AM	8:25:00 AM	LivingRLights	Living room Lights			
8:02:00 AM	8:03:00 AM	KitchenSinkH	Kitchen Sink hot water	2.38	0.63	
8:10:00 AM	8:11:00 AM	KitchenSinkH	Kitchen Sink hot water	2.38	0.63	
8:20:00 AM	8:21:00 AM	KitchenSinkH	Kitchen Sink hot water	2.38	0.63	
8:28:00 AM	8:29:00 AM	KitchenSinkH	Kitchen Sink hot water	2.38	0.63	
12:00:00 PM	12:10:00 PM	MBA Tub Cold	Master Bathroom cold water			
12:00:00 PM	1:00:00 PM	MBA Drain	Master Bathroom Drain Tank Solenoid			
4:00:00 PM	11:59:00 PM	Latent Load	Latent Load			
6:00:00 PM	6:06:00 PM	LightMastBath	Master Bathroom Lights			
6:00:00 PM	6:30:00 PM	KitchenLights	Kitchen Lights			
6:00:00 PM	6:30:00 PM	LivingRLights	Living room Lights			
6:00:00 PM	6:30:00 PM	SHchildBDR	Sensible Heat ChildB Dining Room			
6:00:00 PM	6:30:00 PM	SHparentAMBR	Sensible Heat ParentA Master Bedroom			
6:00:00 PM	7:30:00 PM	SHchildABR2	Sensible Heat ChildA Bedroom # 2			
6:00:00 PM	10:00:00 PM	SHparentBDR	Sensible Heat ParentB Dining Room			
6:02:00 PM	7:02:00 PM	CookTops	Cooktops			408.04
6:05:00 PM	6:06:00 PM	SinkMastBathH	Sink event - Master Bathroom hot water	2.38	0.63	
6:07:00 PM	6:19:00 PM	ShowerMastBathH	Shower even - Master Bathroom hot water	33.12	8.75	
6:07:00 PM	6:30:00 PM	LightMastBed	Master Bedroom Lights			
6:08:00 PM	6:18:00 PM	Light2ndBath	2nd Bathroom Lights			
6:10:00 PM	8:30:00 PM	LivingRATV	Living room - Active TV			

6:10:00 PM		MicrowaveON	Turn Microwave ON			
6:20:00 PM	6:21:00 PM	SinkMastBathH	Sink event - Master Bathroom hot water	2.38	0.63	
6:20:00 PM	10:00:00 PM	Light2ndBed	2nd Bedroom Lights			
6:25:00 PM	6:26:00 PM	KitchenSinkH	Kitchen Sink hot water	2.38	0.63	
6:30:00 PM	7:30:00 PM	SHchildBBR3	Sensible Heat ChildB Bedroom # 3			
6:30:00 PM	8:15:00 PM	ComputerACB	Bedroom # 3 - Laptop			
6:30:00 PM	8:30:00 PM	Light3rdBed	3rd Bedroom Lights			
6:30:00 PM	10:00:00 PM	ComputerACA	Bedroom # 2 - Laptop			
6:30:00 PM	10:30:00 PM	DesktopCA	Desktop PC and Monitor			342.60
6:30:00 PM	10:30:00 PM	SHparentAK	Sensible Heat ParentA Kitchen			
6:30:00 PM		WashingM	Clothes Washer			
6:40:00 PM	6:41:00 PM	KitchenSinkH	Kitchen Sink hot water	2.38	0.63	
6:40:00 PM	6:47:04 PM	CanOpener	Can Opener			
6:40:00 PM	6:50:00 PM	HandMixer	Hand Mixer			5.50
6:45:00 PM	6:46:00 PM	KitchenSinkH	Kitchen Sink hot water	2.38	0.63	
6:55:00 PM	6:56:00 PM	KitchenSinkH	Kitchen Sink hot water	2.38	0.63	
7:00:00 PM	7:15:00 PM	KitchenLights	Kitchen Lights			
7:00:00 PM	7:15:00 PM	LivingRLights	Living room Lights			
7:05:00 PM	7:06:00 PM	KitchenSinkH	Kitchen Sink hot water	2.38	0.63	
7:10:00 PM	7:11:00 PM	KitchenSinkH	Kitchen Sink hot water	2.38	0.63	
7:15:00 PM	8:30:00 PM	DiningRLights	Dining Room Lights			
7:30:00 PM	8:15:00 PM	KitchenLights	Kitchen Lights			
7:30:00 PM	8:15:00 PM	SHchildALR	Sensible Heat ChildA Living Room			
7:30:00 PM	8:15:00 PM	SHchildBDR	Sensible Heat ChildB Dining Room			
7:30:00 PM	10:30:00 PM	LivingRLights	Living room Lights			
7:35:00 PM	7:36:00 PM	KitchenSinkH	Kitchen Sink hot water	2.38	0.63	
8:00:00 PM	8:01:00 PM	KitchenSinkH	Kitchen Sink hot water	2.38	0.63	
8:11:00 PM	8:12:00 PM	KitchenSinkH	Kitchen Sink hot water	2.38	0.63	
8:13:00 PM	8:25:00 PM	MBA Tub	Master Bathroom Tub	113.56	30.00	
8:15:00 PM	8:25:00 PM	Light2ndBath	2nd Bathroom Lights			
8:15:00 PM	11:59:00 PM	SHchildABR2	Sensible Heat ChildA Bedroom # 2			
8:15:00 PM	11:59:00 PM	SHchildBBR3	Sensible Heat ChildB Bedroom # 3			
8:25:00 PM		WashingM	Clothes Washer			
8:26:00 PM	8:27:00 PM	KitchenSinkH	Kitchen Sink hot water	2.38	0.63	
8:28:00 PM	8:29:00 PM	SinkMastBathH	Sink event - Master Bathroom hot water	2.38	0.63	
8:29:00 PM		Dishwasher	Dishwasher			
8:30:00 PM	10:30:00 PM	KitchenLights	Kitchen Lights			
8:30:00 PM		ClothDryer	Clothes Dryer			
8:35:00 PM	8:36:00 PM	KitchenSinkH	Kitchen Sink hot water	2.38	0.63	
8:50:00 PM	8:51:00 PM	KitchenSinkH	Kitchen Sink hot water	2.38	0.63	

9:05:00 PM	9:06:00 PM	KitchenSinkH	Kitchen Sink hot water	2.38	0.63	
9:10:00 PM	9:11:00 PM	KitchenSinkH	Kitchen Sink hot water	2.38	0.63	
9:35:00 PM	9:36:00 PM	KitchenSinkH	Kitchen Sink hot water	2.38	0.63	
9:38:00 PM	9:43:00 PM	Light1stBath	1st Bathroom Lights			
9:40:00 PM	9:41:00 PM	KitchenSinkH	Kitchen Sink hot water	2.38	0.63	
9:45:00 PM	10:00:00 PM	Light2ndBath	2nd Bathroom Lights			
9:55:00 PM	9:56:00 PM	SinkMastBathH	Sink event - Master Bathroom hot water	2.38	0.63	
10:00:00 PM	10:15:00 PM	HeatingPad	Heating Pad			
10:00:00 PM	11:00:00 PM	LightMastBed	Master Bedroom Lights			
10:00:00 PM	11:30:00 PM	BlueRayM	Master Bedroom - BlueRay			
10:00:00 PM	11:30:00 PM	MastBedATV	Master Bedroom Active TV			
10:00:00 PM	11:59:00 PM	PortableFan	Portable Fan			
10:00:00 PM	11:59:00 PM	SHparentBMBR	Sensible Heat ParentB Master Bedroom			
10:05:00 PM	10:06:00 PM	KitchenSinkH	Kitchen Sink hot water	2.38	0.63	
10:05:00 PM	10:25:00 PM	LightMastBath	Master Bathroom Lights			
10:10:00 PM	10:11:00 PM	KitchenSinkH	Kitchen Sink hot water	2.38	0.63	
10:20:00 PM	10:21:00 PM	SinkMastBathH	Sink event - Master Bathroom hot water	2.38	0.63	
10:30:00 PM	10:40:00 PM	LightMastBath	Master Bathroom Lights			
10:30:00 PM	11:59:00 PM	SHparentAMBR	Sensible Heat ParentA Master Bedroom			
10:40:00 PM	10:41:00 PM	SinkMastBathH	Sink event - Master Bathroom hot water	2.38	0.63	

Thursdays

Start time	End time	FixtureID	Description	Target Volume / Draw (liters)	Target Volume / Draw (gallons)	Target Energy / Event (Wh)
12:05:00 AM	6:00:00 AM	SHparentBMBR	Sensible Heat ParentB Master Bedroom			
12:05:00 AM	6:30:00 AM	SHparentAMBR	Sensible Heat ParentA Master Bedroom			
12:05:00 AM	7:00:00 AM	SHchildABR2	Sensible Heat ChildA Bedroom # 2			
12:05:00 AM	7:00:00 AM	SHchildBBR3	Sensible Heat ChildB Bedroom # 3			
12:05:00 AM	8:30:00 AM	Latent Load	Latent Load			
3:31:00 AM	3:34:00 AM	LightMastBath	Master Bathroom Lights			
3:33:00 AM	3:34:00 AM	SinkMastBathH	Sink event - Master Bathroom hot water	2.38	0.63	
4:00:00 AM	4:03:00 AM	LightMastBath	Master Bathroom Lights			
4:02:00 AM	4:03:00 AM	SinkMastBathH	Sink event - Master Bathroom hot water	2.38	0.63	
6:00:00 AM	6:01:00 AM	KitchenSinkH	Kitchen Sink hot water	2.38	0.63	
6:00:00 AM	6:15:00 AM	LightMastBath	Master Bathroom Lights			
6:00:00 AM	6:30:00 AM	SHparentBDR	Sensible Heat ParentB Dining Room			
6:00:00 AM	8:25:00 AM	KitchenLights	Kitchen Lights			
6:00:00 AM		MicrowaveCLR	Clear Microwave (equivalent to opening and closing the door)			
6:01:00 AM		MicrowaveON	Turn Microwave ON			
6:02:00 AM	6:03:00 AM	KitchenSinkH	Kitchen Sink hot water	2.38	0.63	
6:04:00 AM	6:14:00 AM	ShowerMastBathH	Shower even - Master Bathroom hot water	33.12	8.75	
6:05:00 AM	6:15:00 AM	Blender	Blender			19.20
6:10:00 AM	6:40:00 AM	CoffeeMaker	Coffee Maker			168.12
6:15:00 AM	6:16:00 AM	SinkMastBathH	Sink event - Master Bathroom hot water	2.38	0.63	
6:15:00 AM	6:25:00 AM	ToasterOven	Toaster Oven			88.80
6:15:00 AM	6:30:00 AM	LightMastBed	Master Bedroom Lights			
6:15:00 AM	7:00:00 AM	MastBedATV	Master Bedroom Active TV			
6:20:00 AM	6:35:00 AM	Toaster	Toaster			126.00
6:25:00 AM	6:26:00 AM	KitchenSinkH	Kitchen Sink hot water	2.38	0.63	
6:30:00 AM	6:40:00 AM	ShowerMastBathH	Shower even - Master Bathroom hot water	33.12	8.75	
6:30:00 AM	6:45:00 AM	LightMastBath	Master Bathroom Lights			
6:30:00 AM	6:50:00 AM	Light2ndBath	2nd Bathroom Lights			
6:30:00 AM	7:00:00 AM	SHparentBMBR	Sensible Heat ParentB Master Bedroom			
6:30:00 AM	8:25:00 AM	LivingRATV	Living room - Active TV			
6:30:00 AM	8:30:00 AM	SHparentAK	Sensible Heat ParentA Kitchen			

6:40:00 AM	6:50:00 AM	HairDryer	Hair Dryer			115.70
6:40:00 AM	7:00:00 AM	Light2ndBed	2nd Bedroom Lights			
6:41:00 AM	6:42:00 AM	SinkMastBathH	Sink event - Master Bathroom hot water	2.38	0.63	
6:43:00 AM	6:44:00 AM	KitchenSinkH	Kitchen Sink hot water	2.38	0.63	
6:45:00 AM	6:55:00 AM	ShowerMastBathH	Shower even - Master Bathroom hot water	33.12	8.75	
6:45:00 AM	7:00:00 AM	LightMastBed	Master Bedroom Lights			
6:50:00 AM	7:00:00 AM	Light3rdBed	3rd Bedroom Lights			
6:56:00 AM	6:57:00 AM	SinkMastBathH	Sink event - Master Bathroom hot water	2.38	0.63	
6:58:00 AM	6:59:00 AM	SinkMastBathH	Sink event - Master Bathroom hot water	2.38	0.63	
7:00:00 AM	8:00:00 AM	DiningRLights	Dining Room Lights			
7:00:00 AM	8:00:00 AM	SHchildALR	Sensible Heat ChildA Living Room			
7:00:00 AM	8:00:00 AM	SHchildBDR	Sensible Heat ChildB Dining Room			
7:00:00 AM	8:30:00 AM	SHparentBDR	Sensible Heat ParentB Dining Room			
7:05:00 AM	7:06:00 AM	KitchenSinkH	Kitchen Sink hot water	2.38	0.63	
7:30:00 AM	7:31:00 AM	KitchenSinkH	Kitchen Sink hot water	2.38	0.63	
7:35:00 AM	7:36:00 AM	KitchenSinkH	Kitchen Sink hot water	2.38	0.63	
8:00:00 AM	8:25:00 AM	LivingRLights	Living room Lights			
8:02:00 AM	8:03:00 AM	KitchenSinkH	Kitchen Sink hot water	2.38	0.63	
8:10:00 AM	8:11:00 AM	KitchenSinkH	Kitchen Sink hot water	2.38	0.63	
8:20:00 AM	8:21:00 AM	KitchenSinkH	Kitchen Sink hot water	2.38	0.63	
8:26:00 AM	8:27:00 AM	KitchenSinkH	Kitchen Sink hot water	2.38	0.63	
4:00:00 PM	4:30:00 PM	SHchildALR	Sensible Heat ChildA Living Room			
4:00:00 PM	11:59:00 PM	Latent Load	Latent Load			
4:01:00 PM	4:06:00 PM	Light1stBath	1st Bathroom Lights			
4:04:00 PM	4:05:00 PM	KitchenSinkH	Kitchen Sink hot water	2.38	0.63	
4:10:00 PM	4:11:00 PM	KitchenSinkH	Kitchen Sink hot water	2.38	0.63	
4:25:00 PM	4:26:00 PM	KitchenSinkH	Kitchen Sink hot water	2.38	0.63	
4:30:00 PM	6:00:00 PM	Light2ndBed	2nd Bedroom Lights			
4:30:00 PM	6:00:00 PM	SHchildABR2	Sensible Heat ChildA Bedroom # 2			
6:00:00 PM	6:30:00 PM	KitchenLights	Kitchen Lights			
6:00:00 PM	6:30:00 PM	LivingRLights	Living room Lights			
6:00:00 PM	6:30:00 PM	SHchildALR	Sensible Heat ChildA Living Room			
6:00:00 PM	6:30:00 PM	SHchildBDR	Sensible Heat ChildB Dining Room			
6:00:00 PM	6:30:00 PM	SHparentAMBR	Sensible Heat ParentA Master Bedroom			
6:00:00 PM	10:00:00 PM	SHparentBDR	Sensible Heat ParentB Dining Room			
6:01:00 PM	6:06:00 PM	LightMastBath	Master Bathroom Lights			
6:02:00 PM	6:06:00 PM	Light1stBath	1st Bathroom Lights			
6:02:00 PM	6:08:49 PM	RangeHood	Range-hood			
6:02:00 PM	7:02:00 PM	CookTops	Cooktops			408.04

44

6:05:00 PM	6:06:00 PM	SinkMastBathH	Sink event - Master Bathroom hot water	2.38	0.63	
6:07:00 PM	6:30:00 PM	LightMastBed	Master Bedroom Lights			
6:09:00 PM		MicrowaveCLR	Clear Microwave (equivalent to opening and closing the door)			
6:10:00 PM	8:30:00 PM	LivingRATV	Living room - Active TV			
6:10:00 PM		MicrowaveON	Turn Microwave ON			
6:30:00 PM	7:30:00 PM	SHchildABR2	Sensible Heat ChildA Bedroom # 2			
6:30:00 PM	7:30:00 PM	SHchildBBR3	Sensible Heat ChildB Bedroom # 3			
6:30:00 PM	8:15:00 PM	ComputerACB	Bedroom # 3 - Laptop			
6:30:00 PM	8:30:00 PM	Light3rdBed	3rd Bedroom Lights			
6:30:00 PM	10:00:00 PM	ComputerACA	Bedroom # 2 - Laptop			
6:30:00 PM	10:00:00 PM	Light2ndBed	2nd Bedroom Lights			
6:30:00 PM	10:30:00 PM	DesktopCA	Desktop PC and Monitor			342.60
6:30:00 PM	10:30:00 PM	SHparentAK	Sensible Heat ParentA Kitchen			
6:40:00 PM	6:41:00 PM	KitchenSinkH	Kitchen Sink hot water	2.38	0.63	
6:40:00 PM	6:47:04 PM	CanOpener	Can Opener			
6:40:00 PM	6:50:00 PM	HandMixer	Hand Mixer			5.50
6:45:00 PM	6:46:00 PM	KitchenSinkH	Kitchen Sink hot water	2.38	0.63	
6:55:00 PM	6:56:00 PM	KitchenSinkH	Kitchen Sink hot water	2.38	0.63	
7:00:00 PM	7:15:00 PM	KitchenLights	Kitchen Lights			
7:00:00 PM	7:15:00 PM	LivingRLights	Living room Lights			
7:10:00 PM	7:11:00 PM	KitchenSinkH	Kitchen Sink hot water	2.38	0.63	
7:15:00 PM	8:30:00 PM	DiningRLights	Dining Room Lights			
7:30:00 PM	8:15:00 PM	KitchenLights	Kitchen Lights			
7:30:00 PM	8:15:00 PM	SHchildALR	Sensible Heat ChildA Living Room			
7:30:00 PM	8:15:00 PM	SHchildBDR	Sensible Heat ChildB Dining Room			
7:30:00 PM	9:30:00 PM	LivingRLights	Living room Lights			
7:35:00 PM	7:36:00 PM	KitchenSinkH	Kitchen Sink hot water	2.38	0.63	
8:00:00 PM	8:01:00 PM	KitchenSinkH	Kitchen Sink hot water	2.38	0.63	
8:15:00 PM	8:16:00 PM	KitchenSinkH	Kitchen Sink hot water	2.38	0.63	
8:15:00 PM	8:25:00 PM	Light2ndBath	2nd Bathroom Lights			
8:15:00 PM	11:59:00 PM	SHchildABR2	Sensible Heat ChildA Bedroom # 2			
8:15:00 PM	11:59:00 PM	SHchildBBR3	Sensible Heat ChildB Bedroom # 3			
8:20:00 PM	8:21:00 PM	SinkMastBathH	Sink event - Master Bathroom hot water	2.38	0.63	
8:25:00 PM	8:26:00 PM	KitchenSinkH	Kitchen Sink hot water	2.38	0.63	
8:30:00 PM	10:30:00 PM	KitchenLights	Kitchen Lights			
8:35:00 PM	8:36:00 PM	KitchenSinkH	Kitchen Sink hot water	2.38	0.63	
8:50:00 PM	8:51:00 PM	KitchenSinkH	Kitchen Sink hot water	2.38	0.63	
9:05:00 PM	9:06:00 PM	KitchenSinkH	Kitchen Sink hot water	2.38	0.63	
9:10:00 PM	9:11:00 PM	KitchenSinkH	Kitchen Sink hot water	2.38	0.63	
9:30:00 PM	9:31:00 PM	KitchenSinkH	Kitchen Sink hot water	2.38	0.63	

9:40:00 PM	9:41:00 PM	KitchenSinkH	Kitchen Sink hot water	2.38	0.63	
9:43:00 PM	9:58:00 PM	ShowerMastBathH	Shower even - Master Bathroom hot water	53.00	14.00	
9:45:00 PM	10:00:00 PM	Light2ndBath	2nd Bathroom Lights			
9:59:00 PM	10:00:00 PM	SinkMastBathH	Sink event - Master Bathroom hot water	2.38	0.63	
10:00:00 PM	12:00:00 AM	PortableFan	Portable Fan			
10:00:00 PM	10:10:00 PM	HeatingPad	Heating Pad			
10:00:00 PM	10:30:00 PM	LivingRLights	Living room Lights			
10:00:00 PM	11:00:00 PM	LightMastBed	Master Bedroom Lights			
10:00:00 PM	11:59:00 PM	SHparentBMBR	Sensible Heat ParentB Master Bedroom			
10:05:00 PM	10:06:00 PM	KitchenSinkH	Kitchen Sink hot water	2.38	0.63	
10:05:00 PM	10:25:00 PM	LightMastBath	Master Bathroom Lights			
10:20:00 PM	10:21:00 PM	SinkMastBathH	Sink event - Master Bathroom hot water	2.38	0.63	
10:30:00 PM	10:40:00 PM	LightMastBath	Master Bathroom Lights			
10:30:00 PM	11:59:00 PM	SHparentAMBR	Sensible Heat ParentA Master Bedroom			
10:35:00 PM	10:36:00 PM	SinkMastBathH	Sink event - Master Bathroom hot water	2.38	0.63	

Fridays

Start time	End time	FixtureID	Description	Target Volume / Draw (liters)	Target Volume / Draw (gallons)	Target Energy / Event (Wh)
12:05:00 AM	6:00:00 AM	SHparentBMBR	Sensible Heat ParentB Master Bedroom			
12:05:00 AM	6:30:00 AM	SHparentAMBR	Sensible Heat ParentA Master Bedroom			
12:05:00 AM	7:00:00 AM	SHchildABR2	Sensible Heat ChildA Bedroom # 2			
12:05:00 AM	7:00:00 AM	SHchildBBR3	Sensible Heat ChildB Bedroom # 3			
12:05:00 AM	8:30:00 AM	Latent Load	Latent Load			
3:31:00 AM	3:34:00 AM	LightMastBath	Master Bathroom Lights			
3:33:00 AM	3:34:00 AM	SinkMastBathH	Sink event - Master Bathroom hot water	2.38	0.63	
4:00:00 AM	4:03:00 AM	LightMastBath	Master Bathroom Lights			
4:02:00 AM	4:03:00 AM	SinkMastBathH	Sink event - Master Bathroom hot water	2.38	0.63	
6:00:00 AM	6:15:00 AM	LightMastBath	Master Bathroom Lights			
6:00:00 AM	6:30:00 AM	SHparentBDR	Sensible Heat ParentB Dining Room			
6:00:00 AM	8:25:00 AM	KitchenLights	Kitchen Lights			
6:00:00 AM		MicrowaveCLR	Clear Microwave (equivalent to opening and closing the door)			
6:01:00 AM		MicrowaveON	Turn Microwave ON			
6:04:00 AM	6:05:00 AM	KitchenSinkH	Kitchen Sink hot water	2.38	0.63	
6:05:00 AM	6:15:00 AM	Blender	Blender			19.20
6:06:00 AM	6:17:00 AM	ShowerMastBathH	Shower even - Master Bathroom hot water	33.12	8.75	
6:10:00 AM	6:40:00 AM	CoffeeMaker	Coffee Maker			168.12
6:15:00 AM	6:25:00 AM	ToasterOven	Toaster Oven			88.80
6:15:00 AM	6:30:00 AM	LightMastBed	Master Bedroom Lights			
6:15:00 AM	7:00:00 AM	MastBedATV	Master Bedroom Active TV			
6:18:00 AM	6:19:00 AM	SinkMastBathH	Sink event - Master Bathroom hot water	2.38	0.63	
6:20:00 AM	6:35:00 AM	Toaster	Toaster			126.00
6:27:00 AM	6:28:00 AM	SinkMastBathH	Sink event - Master Bathroom hot water	2.38	0.63	
6:29:00 AM	6:30:00 AM	KitchenSinkH	Kitchen Sink hot water	2.38	0.63	
6:30:00 AM	6:45:00 AM	LightMastBath	Master Bathroom Lights			
6:30:00 AM	6:50:00 AM	Light2ndBath	2nd Bathroom Lights			
6:30:00 AM	7:00:00 AM	SHparentBMBR	Sensible Heat ParentB Master Bedroom			
6:30:00 AM	8:30:00 AM	SHparentAK	Sensible Heat ParentA Kitchen			
6:31:00 AM	6:41:00 AM	ShowerMastBathH	Shower even - Master Bathroom hot water	33.12	8.75	
6:40:00 AM	6:50:00 AM	HairDryer	Hair Dryer			115.70

6:40:00 AM	7:00:00 AM	Light2ndBed	2nd Bedroom Lights			
6:42:00 AM	6:43:00 AM	SinkMastBathH	Sink event - Master Bathroom hot water	2.38	0.63	
6:44:00 AM	6:54:00 AM	ShowerMastBathH	Shower even - Master Bathroom hot water	33.12	8.75	
6:45:00 AM	7:00:00 AM	LightMastBed	Master Bedroom Lights			
6:50:00 AM	7:00:00 AM	Light3rdBed	3rd Bedroom Lights			
6:55:00 AM	6:56:00 AM	SinkMastBathH	Sink event - Master Bathroom hot water	2.38	0.63	
7:00:00 AM	8:00:00 AM	DiningRLights	Dining Room Lights			
7:00:00 AM	8:00:00 AM	SHchildALR	Sensible Heat ChildA Living Room			
7:00:00 AM	8:00:00 AM	SHchildBDR	Sensible Heat ChildB Dining Room			
7:00:00 AM	8:30:00 AM	SHparentBDR	Sensible Heat ParentB Dining Room			
7:05:00 AM	7:06:00 AM	KitchenSinkH	Kitchen Sink hot water	2.38	0.63	
7:30:00 AM	7:31:00 AM	KitchenSinkH	Kitchen Sink hot water	2.38	0.63	
8:00:00 AM	8:25:00 AM	LivingRLights	Living room Lights			
8:02:00 AM	8:03:00 AM	KitchenSinkH	Kitchen Sink hot water	2.38	0.63	
8:10:00 AM	8:11:00 AM	KitchenSinkH	Kitchen Sink hot water	2.38	0.63	
8:26:00 AM	8:27:00 AM	KitchenSinkH	Kitchen Sink hot water	2.38	0.63	
8:28:00 AM	8:29:00 AM	KitchenSinkH	Kitchen Sink hot water	2.38	0.63	
4:00:00 PM	11:59:00 PM	Latent Load	Latent Load			
6:00:00 PM	6:30:00 PM	KitchenLights	Kitchen Lights			
6:00:00 PM	6:30:00 PM	SHchildBDR	Sensible Heat ChildB Dining Room			
6:00:00 PM	6:30:00 PM	SHparentAMBR	Sensible Heat ParentA Master Bedroom			
6:00:00 PM	7:30:00 PM	SHchildABR2	Sensible Heat ChildA Bedroom # 2			
6:00:00 PM	11:30:00 PM	SHparentBDR	Sensible Heat ParentB Dining Room			
6:01:00 PM	6:06:00 PM	LightMastBath	Master Bathroom Lights			
6:01:00 PM	6:30:00 PM	LivingRLights	Living room Lights			
6:02:00 PM	7:02:00 PM	CookTops	Cooktops			408.04
6:05:00 PM	6:06:00 PM	SinkMastBathH	Sink event - Master Bathroom hot water	2.38	0.63	
6:07:00 PM	6:30:00 PM	LightMastBed	Master Bedroom Lights			
6:08:00 PM	6:09:00 PM	KitchenSinkH	Kitchen Sink hot water	2.38	0.63	
6:08:00 PM	6:18:00 PM	Light2ndBath	2nd Bathroom Lights			
6:09:00 PM		MicrowaveCLR	Clear Microwave (equivalent to opening and closing the door)			
6:10:00 PM	6:25:00 PM	ShowerMastBathH	Shower even - Master Bathroom hot water	53.00	14.00	
6:10:00 PM	10:25:00 PM	LivingRATV	Living room - Active TV			
6:10:00 PM		MicrowaveON	Turn Microwave ON			
6:20:00 PM	9:00:00 PM	Light2ndBed	2nd Bedroom Lights			
6:26:00 PM	6:27:00 PM	KitchenSinkH	Kitchen Sink hot water	2.38	0.63	
6:28:00 PM	6:29:00 PM	SinkMastBathH	Sink event - Master Bathroom hot water	2.38	0.63	

6:30:00 PM	7:30:00 PM	SHchildBBR3	Sensible Heat ChildB Bedroom # 3			
6:30:00 PM	9:00:00 PM	ComputerACA	Bedroom # 2 - Laptop			
6:30:00 PM	9:00:00 PM	ComputerACB	Bedroom # 3 - Laptop			
6:30:00 PM	9:00:00 PM	Light3rdBed	3rd Bedroom Lights			
6:30:00 PM	10:30:00 PM	DesktopCA	Desktop PC and Monitor			342.60
6:30:00 PM	11:30:00 PM	SHparentAK	Sensible Heat ParentA Kitchen			
6:40:00 PM	6:47:04 PM	CanOpener	Can Opener			
6:40:00 PM	6:50:00 PM	HandMixer	Hand Mixer			5.50
6:55:00 PM	6:56:00 PM	KitchenSinkH	Kitchen Sink hot water	2.38	0.63	
7:00:00 PM	7:15:00 PM	KitchenLights	Kitchen Lights			
7:00:00 PM	7:15:00 PM	LivingRLights	Living room Lights			
7:05:00 PM	7:06:00 PM	KitchenSinkH	Kitchen Sink hot water	2.38	0.63	
7:10:00 PM	7:11:00 PM	KitchenSinkH	Kitchen Sink hot water	2.38	0.63	
7:15:00 PM	9:00:00 PM	DiningRLights	Dining Room Lights			
7:30:00 PM	8:15:00 PM	KitchenLights	Kitchen Lights			
7:30:00 PM	8:15:00 PM	SHchildALR	Sensible Heat ChildA Living Room			
7:30:00 PM	8:15:00 PM	SHchildBDR	Sensible Heat ChildB Dining Room			
7:30:00 PM	11:30:00 PM	LivingRLights	Living room Lights			
7:35:00 PM	7:36:00 PM	KitchenSinkH	Kitchen Sink hot water	2.38	0.63	
8:00:00 PM	9:00:00 PM	Light2ndBath	2nd Bathroom Lights			
8:00:00 PM	8:01:00 PM	KitchenSinkH	Kitchen Sink hot water	2.38	0.63	
8:15:00 PM	9:00:00 PM	SHchildABR2	Sensible Heat ChildA Bedroom # 2			
8:15:00 PM	9:00:00 PM	SHchildBBR3	Sensible Heat ChildB Bedroom # 3			
8:15:00 PM		Dishwasher	Dishwasher			
8:25:00 PM	8:26:00 PM	KitchenSinkH	Kitchen Sink hot water	2.38	0.63	
9:00:00 PM	10:00:00 PM	SHchildBDR	Sensible Heat ChildB Dining Room			
9:00:00 PM	10:00:00 PM	VideoAGame	Video Game Active (Turned ON)			
9:00:00 PM	11:00:00 PM	SHchildALR	Sensible Heat ChildA Living Room			
9:00:00 PM	11:30:00 PM	KitchenLights	Kitchen Lights			
9:05:00 PM	9:06:00 PM	KitchenSinkH	Kitchen Sink hot water	2.38	0.63	
9:10:00 PM	9:11:00 PM	KitchenSinkH	Kitchen Sink hot water	2.38	0.63	
9:30:00 PM	9:31:00 PM	KitchenSinkH	Kitchen Sink hot water	2.38	0.63	
9:40:00 PM	9:41:00 PM	KitchenSinkH	Kitchen Sink hot water	2.38	0.63	
10:00:00 PM	10:01:00 PM	KitchenSinkH	Kitchen Sink hot water	2.38	0.63	
10:00:00 PM	10:15:00 PM	Light2ndBath	2nd Bathroom Lights			
10:00:00 PM	11:59:00 PM	SHchildBBR3	Sensible Heat ChildB Bedroom # 3			
10:05:00 PM	10:06:00 PM	KitchenSinkH	Kitchen Sink hot water	2.38	0.63	
10:10:00 PM	10:11:00 PM	SinkMastBathH	Sink event - Master Bathroom hot water	2.38	0.63	
10:15:00 PM	10:16:00 PM	KitchenSinkH	Kitchen Sink hot water	2.38	0.63	

10:18:00 PM	10:22:00 PM	Light1stBath	1st Bathroom Lights			
10:20:00 PM	10:21:00 PM	KitchenSinkH	Kitchen Sink hot water	2.38	0.63	
10:30:00 PM	10:31:00 PM	KitchenSinkH	Kitchen Sink hot water	2.38	0.63	
10:40:00 PM	10:41:00 PM	KitchenSinkH	Kitchen Sink hot water	2.38	0.63	
11:00:00 PM	11:15:00 PM	Light2ndBath	2nd Bathroom Lights			
11:00:00 PM	11:59:00 PM	SHchildABR2	Sensible Heat ChildA Bedroom # 2			
11:10:00 PM	11:11:00 PM	SinkMastBathH	Sink event - Master Bathroom hot water	2.38	0.63	
11:18:00 PM	11:22:00 PM	Light1stBath	1st Bathroom Lights			
11:20:00 PM	11:21:00 PM	KitchenSinkH	Kitchen Sink hot water	2.38	0.63	
11:25:00 PM	11:26:00 PM	KitchenSinkH	Kitchen Sink hot water	2.38	0.63	
11:30:00 PM	11:45:00 PM	HeatingPad	Heating Pad			
11:30:00 PM	11:45:00 PM	LightMastBath	Master Bathroom Lights			
11:30:00 PM	11:59:00 PM	LightMastBed	Master Bedroom Lights			
11:30:00 PM	11:59:00 PM	PortableFan	Portable Fan			
11:30:00 PM	11:59:00 PM	SHparentAMBR	Sensible Heat ParentA Master Bedroom			
11:30:00 PM	11:59:00 PM	SHparentBMBR	Sensible Heat ParentB Master Bedroom			
11:40:00 PM	11:41:00 PM	SinkMastBathH	Sink event - Master Bathroom hot water	2.38	0.63	
11:50:00 PM	11:51:00 PM	SinkMastBathH	Sink event - Master Bathroom hot water	2.38	0.63	

Saturdays

Start time	End time	FixtureID	Description	Target Volume / Draw (liters)	Target Volume / Draw (gallons)	Target Energy / Event (Wh)
12:05:00 AM	8:15:00 AM	SHparentAMBR	Sensible Heat ParentA Master Bedroom			
12:05:00 AM	8:15:00 AM	SHparentBMBR	Sensible Heat ParentB Master Bedroom			
12:05:00 AM	9:00:00 AM	SHchildBBR3	Sensible Heat ChildB Bedroom # 3			
12:05:00 AM	9:30:00 AM	SHchildABR2	Sensible Heat ChildA Bedroom # 2			
12:05:00 AM	11:59:00 AM	Latent Load	Latent Load			
8:00:00 AM	8:15:00 AM	LightMastBath	Master Bathroom Lights			
8:04:00 AM	8:05:00 AM	SinkMastBathH	Sink event - Master Bathroom hot water	2.38	0.63	
8:10:00 AM	8:11:00 AM	SinkMastBathH	Sink event - Master Bathroom hot water	2.38	0.63	
8:14:00 AM	8:15:00 AM	KitchenSinkH	Kitchen Sink hot water	2.38	0.63	
8:15:00 AM	9:15:00 AM	CookTops	Cooktops			408.04
8:15:00 AM	10:00:00 AM	KitchenLights	Kitchen Lights			
8:15:00 AM	10:00:00 AM	SHparentAK	Sensible Heat ParentA Kitchen			
8:15:00 AM	10:00:00 AM	SHparentBDR	Sensible Heat ParentB Dining Room			
8:15:00 AM	11:15:00 AM	Oven	Oven			1946.79
8:16:00 AM	8:17:00 AM	SinkMastBathH	Sink event - Master Bathroom hot water	2.38	0.63	
8:20:00 AM	8:21:00 AM	KitchenSinkH	Kitchen Sink hot water	2.38	0.63	
8:30:00 AM	8:31:00 AM	KitchenSinkH	Kitchen Sink hot water	2.38	0.63	
8:30:00 AM	11:00:00 AM	LivingRATV	Living room - Active TV			
8:30:00 AM	7:00:00 PM	DesktopCA	Desktop PC and Monitor			899.30
8:40:00 AM	8:41:00 AM	KitchenSinkH	Kitchen Sink hot water	2.38	0.63	
8:45:00 AM	9:15:00 AM	Light2ndBath	2nd Bathroom Lights			
8:50:00 AM	8:51:00 AM	SinkMastBathH	Sink event - Master Bathroom hot water	2.38	0.63	
8:59:00 AM	9:11:00 AM	ShowerMastBathH	Shower even - Master Bathroom hot water	33.12	8.75	
9:00:00 AM	10:00:00 AM	SHchildBDR	Sensible Heat ChildB Dining Room			
9:10:00 AM	9:40:00 AM	CoffeeMaker	Coffee Maker			168.12
9:12:00 AM	9:13:00 AM	KitchenSinkH	Kitchen Sink hot water	2.38	0.63	
9:14:00 AM		MicrowaveCLR	Clear Microwave (equivalent to opening and closing the door)			
9:15:00 AM	9:25:00 AM	Blender	Blender			19.20
9:15:00 AM		MicrowaveON	Turn Microwave ON			
9:16:00 AM	9:17:00 AM	SinkMastBathH	Sink event - Master Bathroom hot water	2.38	0.63	
9:20:00 AM	9:30:00 AM	ToasterOven	Toaster Oven			88.74
9:20:00 AM	9:35:00 AM	Toaster	Toaster			126.10
9:30:00 AM	10:00:00 AM	DiningRLights	Dining Room Lights			

9:30:00 AM	11:00:00 AM	SHchildALR	Sensible Heat ChildA Living Room			
9:32:00 AM	9:33:00 AM	KitchenSinkH	Kitchen Sink hot water	2.38	0.63	
9:40:00 AM	9:41:00 AM	KitchenSinkH	Kitchen Sink hot water	2.38	0.63	
9:45:00 AM	8:30:00 PM	SlowCooker	Slow Cooker			153.85
10:00:00 AM	10:10:00 AM	ShowerMastBathH	Shower even - Master Bathroom hot water	33.12	8.75	
10:00:00 AM	10:15:00 AM	LightMastBath	Master Bathroom Lights			
10:00:00 AM	11:00:00 AM	SHparentBMBR	Sensible Heat ParentB Master Bedroom			
10:00:00 AM	11:00:00 AM	VideoAGame	Video Game Active (Turned ON)			
10:09:00 AM	10:19:00 AM	HairDryer	Hair Dryer			115.70
10:11:00 AM	10:12:00 AM	SinkMastBathH	Sink event - Master Bathroom hot water	2.38	0.63	
10:15:00 AM	11:00:00 AM	LightMastBed	Master Bedroom Lights			
10:30:00 AM	10:31:00 AM	KitchenSinkH	Kitchen Sink hot water	2.38	0.63	
11:00:00 AM	1:00:00 PM	ComputerACA	Bedroom # 2 - Laptop			
11:00:00 AM	1:00:00 PM	SHchildABR2	Sensible Heat ChildA Bedroom # 2			
11:00:00 AM	2:30:00 PM	SHparentBDR	Sensible Heat ParentB Dining Room			
11:00:00 AM		WashingM	Clothes Washer			
12:00:00 PM	12:01:00 PM	KitchenSinkH	Kitchen Sink hot water	2.38	0.63	
12:00:00 PM	11:59:00 PM	Latent Load	Latent Load			
12:02:00 PM	1:02:00 PM	CookTops	Cooktops			408.04
12:09:00 PM		MicrowaveCLR	Clear Microwave (equivalent to opening and closing the door)			
12:10:00 PM		MicrowaveON	Turn Microwave ON			
12:30:00 PM	12:31:00 PM	KitchenSinkH	Kitchen Sink hot water	2.38	0.63	
12:30:00 PM	12:45:00 PM	Light2ndBath	2nd Bathroom Lights			
12:30:00 PM	1:00:00 PM	SHchildBBR3	Sensible Heat ChildB Bedroom # 3			
12:30:00 PM	6:00:00 PM	SHparentAK	Sensible Heat ParentA Kitchen			
12:30:00 PM		ClothDryer	Clothes Dryer			
12:32:00 PM	12:43:00 PM	MBA Tub	Master Bathroom Tub	113.56	30.00	
12:44:00 PM	12:45:00 PM	SinkMastBathH	Sink event - Master Bathroom hot water	2.38	0.63	
12:45:00 PM	1:00:00 PM	Light2ndBed	2nd Bedroom Lights			
1:00:00 PM	1:01:00 PM	KitchenSinkH	Kitchen Sink hot water	2.38	0.63	
1:00:00 PM	2:30:00 PM	SHchildALR	Sensible Heat ChildA Living Room			
1:00:00 PM	6:00:00 PM	SHchildBDR	Sensible Heat ChildB Dining Room			
1:05:00 PM	1:06:00 PM	KitchenSinkH	Kitchen Sink hot water	2.38	0.63	
2:00:00 PM	2:01:00 PM	KitchenSinkH	Kitchen Sink hot water	2.38	0.63	
2:10:00 PM	2:11:00 PM	KitchenSinkH	Kitchen Sink hot water	2.38	0.63	
2:20:00 PM	2:21:00 PM	KitchenSinkH	Kitchen Sink hot water	2.38	0.63	
2:30:00 PM		WashingM	Clothes Washer			
3:00:00 PM	4:30:00 PM	BlueRayL	Living room - BlueRay			

3:00:00 PM	4:30:00 PM	LivingRATV	Living room - Active TV			
3:00:00 PM		ClothDryer	Clothes Dryer			
5:00:00 PM	6:00:00 PM	SHchildALR	Sensible Heat ChildA Living Room			
5:00:00 PM	6:00:00 PM	SHparentBDR	Sensible Heat ParentB Dining Room			
5:30:00 PM	5:31:00 PM	KitchenSinkH	Kitchen Sink hot water	2.38	0.63	
5:30:00 PM	7:27:00 PM	VacuumC	Vacuum Cleaner			1075.70
6:00:00 PM	6:45:00 PM	Iron	Iron			506.70
6:00:00 PM	7:00:00 PM	LightMastBed	Master Bedroom Lights			
6:00:00 PM	7:00:00 PM	SHparentBMBR	Sensible Heat ParentB Master Bedroom			
6:00:00 PM	7:30:00 PM	ComputerACA	Bedroom # 2 - Laptop			
6:00:00 PM	7:30:00 PM	ComputerACB	Bedroom # 3 - Laptop			
6:00:00 PM	7:30:00 PM	Light2ndBed	2nd Bedroom Lights			
6:00:00 PM	7:30:00 PM	Light3rdBed	3rd Bedroom Lights			
6:00:00 PM	7:30:00 PM	SHchildABR2	Sensible Heat ChildA Bedroom # 2			
6:00:00 PM	8:30:00 PM	SHchildBBR3	Sensible Heat ChildB Bedroom # 3			
7:00:00 PM	7:01:00 PM	KitchenSinkH	Kitchen Sink hot water	2.38	0.63	
7:00:00 PM	8:00:00 PM	CookTops	Cooktops			408.04
7:00:00 PM	10:00:00 PM	Oven	Oven			1946.79
7:00:00 PM	11:30:00 PM	KitchenLights	Kitchen Lights			
7:00:00 PM	11:30:00 PM	SHparentBDR	Sensible Heat ParentB Dining Room			
7:05:00 PM	7:12:05 PM	CanOpener	Can Opener			
7:05:00 PM	7:15:00 PM	HandMixer	Hand Mixer			5.50
7:20:00 PM	7:21:00 PM	KitchenSinkH	Kitchen Sink hot water	2.38	0.63	
8:00:00 PM	8:01:00 PM	KitchenSinkH	Kitchen Sink hot water	2.38	0.63	
8:00:00 PM	8:15:00 PM	LightMastBath	Master Bathroom Lights			
8:00:00 PM	8:30:00 PM	DiningRLights	Dining Room Lights			
8:00:00 PM	8:30:00 PM	SHparentAMBR	Sensible Heat ParentA Master Bedroom			
8:00:00 PM	11:30:00 PM	LivingRATV	Living room - Active TV			
8:05:00 PM	8:17:00 PM	ShowerMastBathH	Shower even - Master Bathroom hot water	33.12	8.75	
8:15:00 PM	8:30:00 PM	LightMastBed	Master Bedroom Lights			
8:30:00 PM	10:00:00 PM	SHchildBDR	Sensible Heat ChildB Dining Room			
8:30:00 PM	11:00:00 PM	SHchildALR	Sensible Heat ChildA Living Room			
8:30:00 PM	11:30:00 PM	SHparentAK	Sensible Heat ParentA Kitchen			
9:00:00 PM	11:30:00 PM	LivingRLights	Living room Lights			
9:00:00 PM		Dishwasher	Dishwasher			
9:10:00 PM	9:11:00 PM	KitchenSinkH	Kitchen Sink hot water	2.38	0.63	
9:30:00 PM	9:31:00 PM	KitchenSinkH	Kitchen Sink hot water	2.38	0.63	
9:36:00 PM	9:37:00 PM	KitchenSinkH	Kitchen Sink hot water	2.38	0.63	

10:00:00 PM	10:01:00 PM	KitchenSinkH	Kitchen Sink hot water	2.38	0.63	
10:00:00 PM	10:15:00 PM	Light2ndBath	2nd Bathroom Lights			
10:00:00 PM	11:59:00 PM	SHchildBBR3	Sensible Heat ChildB Bedroom # 3			
10:05:00 PM	10:06:00 PM	KitchenSinkH	Kitchen Sink hot water	2.38	0.63	
10:12:00 PM	10:13:00 PM	SinkMastBathH	Sink event - Master Bathroom hot water	2.38	0.63	
10:18:00 PM	10:22:00 PM	Light1stBath	1st Bathroom Lights			
10:20:00 PM	10:21:00 PM	KitchenSinkH	Kitchen Sink hot water	2.38	0.63	
10:30:00 PM	10:31:00 PM	KitchenSinkH	Kitchen Sink hot water	2.38	0.63	
10:40:00 PM	10:41:00 PM	KitchenSinkH	Kitchen Sink hot water	2.38	0.63	
11:00:00 PM	11:15:00 PM	Light2ndBath	2nd Bathroom Lights			
11:00:00 PM	11:59:00 PM	SHchildABR2	Sensible Heat ChildA Bedroom # 2			
11:10:00 PM	11:11:00 PM	SinkMastBathH	Sink event - Master Bathroom hot water	2.38	0.63	
11:18:00 PM	11:22:00 PM	Light1stBath	1st Bathroom Lights			
11:25:00 PM	11:26:00 PM	KitchenSinkH	Kitchen Sink hot water	2.38	0.63	
11:28:00 PM	11:29:00 PM	KitchenSinkH	Kitchen Sink hot water	2.38	0.63	
11:30:00 PM	11:45:00 PM	HeatingPad	Heating Pad			
11:30:00 PM	11:45:00 PM	LightMastBath	Master Bathroom Lights			
11:30:00 PM	11:59:00 PM	LightMastBed	Master Bedroom Lights			
11:30:00 PM	11:59:00 PM	PortableFan	Portable Fan			
11:30:00 PM	11:59:00 PM	SHparentAMBR	Sensible Heat ParentA Master Bedroom			
11:30:00 PM	11:59:00 PM	SHparentBMBR	Sensible Heat ParentB Master Bedroom			
11:40:00 PM	11:41:00 PM	SinkMastBathH	Sink event - Master Bathroom hot water	2.38	0.63	
11:50:00 PM	11:51:00 PM	SinkMastBathH	Sink event - Master Bathroom hot water	2.38	0.63	

Sundays

Start time	End time	FixtureID	Description	Target Volume / Draw (liters)	Target Volume / Draw (gallons)	Target Energy / Event (Wh)
12:05:00 AM	8:15:00 AM	SHparentAMBR	Sensible Heat ParentA Master Bedroom			
12:05:00 AM	8:15:00 AM	SHparentBMBR	Sensible Heat ParentB Master Bedroom			
12:05:00 AM	9:00:00 AM	SHchildABR2	Sensible Heat ChildA Bedroom # 2			
12:05:00 AM	9:00:00 AM	SHchildBBR3	Sensible Heat ChildB Bedroom # 3			
12:05:00 AM	11:59:00 AM	Latent Load	Latent Load			
6:09:00 AM		MicrowaveCLR	Clear Microwave (equivalent to opening and closing the door)			
8:00:00 AM	8:15:00 AM	LightMastBath	Master Bathroom Lights			
8:04:00 AM	8:05:00 AM	SinkMastBathH	Sink event - Master Bathroom hot water	2.38	0.63	
8:10:00 AM	8:11:00 AM	SinkMastBathH	Sink event - Master Bathroom hot water	2.38	0.63	
8:15:00 AM	8:16:00 AM	KitchenSinkH	Kitchen Sink hot water	2.38	0.63	
8:15:00 AM	9:15:00 AM	CookTops	Cooktops			408.04
8:15:00 AM	10:00:00 AM	KitchenLights	Kitchen Lights			
8:15:00 AM	10:00:00 AM	SHparentAK	Sensible Heat ParentA Kitchen			
8:15:00 AM	10:00:00 AM	SHparentBDR	Sensible Heat ParentB Dining Room			
8:15:00 AM	11:15:00 AM	Oven	Oven			1946.79
8:25:00 AM	8:26:00 AM	KitchenSinkH	Kitchen Sink hot water	2.38	0.63	
8:30:00 AM	10:30:00 AM	LivingRATV	Living room - Active TV			
8:30:00 AM	7:00:00 PM	DesktopCA	Desktop PC and Monitor			899.3
8:40:00 AM	8:41:00 AM	KitchenSinkH	Kitchen Sink hot water	2.38	0.63	
8:45:00 AM	9:15:00 AM	Light2ndBath	2nd Bathroom Lights			
8:50:00 AM	8:51:00 AM	SinkMastBathH	Sink event - Master Bathroom hot water	2.38	0.63	
8:55:00 AM	8:56:00 AM	KitchenSinkH	Kitchen Sink hot water	2.38	0.63	
9:00:00 AM	9:11:00 AM	ShowerMastBathH	Shower even - Master Bathroom hot water	33.12	8.75	
9:00:00 AM	9:30:00 AM	SHchildBBR3	Sensible Heat ChildB Bedroom # 3			
9:00:00 AM	11:00:00 AM	SHchildALR	Sensible Heat ChildA Living Room			
9:10:00 AM	9:40:00 AM	CoffeeMaker	Coffee Maker			168.12
9:12:00 AM	9:13:00 AM	KitchenSinkH	Kitchen Sink hot water	2.38	0.63	
9:14:00 AM	9:15:00 AM	SinkMastBathH	Sink event - Master Bathroom hot water	2.38	0.63	
9:14:00 AM		MicrowaveCLR	Clear Microwave (equivalent to opening and closing the door)			
9:15:00 AM	9:25:00 AM	Blender	Blender			19.20
9:15:00 AM		MicrowaveON	Turn Microwave ON			

9:20:00 AM	9:30:00 AM	ToasterOven	Toaster Oven			88.80
9:20:00 AM	9:35:00 AM	Toaster	Toaster			126.00
9:30:00 AM	10:00:00 AM	DiningRLights	Dining Room Lights			
9:30:00 AM	10:00:00 AM	SHchildBDR	Sensible Heat ChildB Dining Room			
9:38:00 AM	9:39:00 AM	KitchenSinkH	Kitchen Sink hot water	2.38	0.63	
10:00:00 AM	10:11:00 AM	ShowerMastBathH	Shower even - Master Bathroom hot water	33.12	8.75	
10:00:00 AM	10:15:00 AM	LightMastBath	Master Bathroom Lights			
10:00:00 AM	11:00:00 AM	SHparentBMBR	Sensible Heat ParentB Master Bedroom			
10:00:00 AM	11:00:00 AM	VideoAGame	Video Game Active (Turned ON)			
10:09:00 AM	10:19:00 AM	HairDryer	Hair Dryer			115.70
10:12:00 AM	10:13:00 AM	SinkMastBathH	Sink event - Master Bathroom hot water	2.38	0.63	
10:15:00 AM	11:00:00 AM	LightMastBed	Master Bedroom Lights			
10:30:00 AM	10:31:00 AM	KitchenSinkH	Kitchen Sink hot water	2.38	0.63	
10:40:00 AM	10:41:00 AM	KitchenSinkH	Kitchen Sink hot water	2.38	0.63	
11:00:00 AM	1:00:00 PM	ComputerACA	Bedroom # 2 - Laptop			
11:00:00 AM	1:00:00 PM	SHchildABR2	Sensible Heat ChildA Bedroom # 2			
11:00:00 AM	2:30:00 PM	SHparentBDR	Sensible Heat ParentB Dining Room			
11:00:00 AM		WashingM	Clothes Washer			
12:00:00 PM	12:01:00 PM	KitchenSinkH	Kitchen Sink hot water	2.38	0.63	
12:00:00 PM	11:59:00 PM	Latent Load	Latent Load			
12:02:00 PM	1:02:00 PM	CookTops	Cooktops			408.04
12:09:00 PM		MicrowaveCLR	Clear Microwave (equivalent to opening and closing the door)			
12:10:00 PM		MicrowaveON	Turn Microwave ON			
12:30:00 PM	12:31:00 PM	KitchenSinkH	Kitchen Sink hot water	2.38	0.63	
12:30:00 PM	1:00:00 PM	SHchildBBR3	Sensible Heat ChildB Bedroom # 3			
12:30:00 PM	2:00:00 PM	SHparentAK	Sensible Heat ParentA Kitchen			
12:30:00 PM		ClothDryer	Clothes Dryer			
12:40:00 PM	12:41:00 PM	KitchenSinkH	Kitchen Sink hot water	2.38	0.63	
1:00:00 PM	1:01:00 PM	KitchenSinkH	Kitchen Sink hot water	2.38	0.63	
1:00:00 PM	2:30:00 PM	SHchildALR	Sensible Heat ChildA Living Room			
1:00:00 PM	6:30:00 PM	SHchildBDR	Sensible Heat ChildB Dining Room			
1:05:00 PM	1:06:00 PM	KitchenSinkH	Kitchen Sink hot water	2.38	0.63	
2:00:00 PM	2:11:00 PM	ShowerMastBathH	Shower even - Master Bathroom hot water	33.12	8.75	
2:00:00 PM	2:15:00 PM	LightMastBath	Master Bathroom Lights			
2:00:00 PM	2:30:00 PM	SHparentAMBR	Sensible Heat ParentA Master Bedroom			
2:12:00 PM	2:13:00 PM	SinkMastBathH	Sink event - Master Bathroom hot water	2.38	0.63	
2:16:00 PM	3:00:00 PM	Iron	Iron			506.70

2:20:00 PM	2:21:00 PM	KitchenSinkH	Kitchen Sink hot water	2.38	0.63	
2:30:00 PM	6:00:00 PM	SHparentAK	Sensible Heat ParentA Kitchen			
2:30:00 PM		WashingM	Clothes Washer			
3:00:00 PM	4:30:00 PM	BlueRayL	Living room - BlueRay			
3:00:00 PM	4:30:00 PM	LivingRATV	Living room - Active TV			
3:00:00 PM		ClothDryer	Clothes Dryer			
5:00:00 PM	6:00:00 PM	SHchildALR	Sensible Heat ChildA Living Room			
5:00:00 PM	10:00:00 PM	SHparentBDR	Sensible Heat ParentB Dining Room			
5:30:00 PM	5:31:00 PM	KitchenSinkH	Kitchen Sink hot water	2.38	0.63	
5:40:00 PM	5:41:00 PM	KitchenSinkH	Kitchen Sink hot water	2.38	0.63	
6:00:00 PM	6:01:00 PM	KitchenSinkH	Kitchen Sink hot water	2.38	0.63	
6:00:00 PM	6:30:00 PM	KitchenLights	Kitchen Lights			
6:00:00 PM	6:30:00 PM	LivingRLights	Living room Lights			
6:00:00 PM	6:30:00 PM	SHparentAMBR	Sensible Heat ParentA Master Bedroom			
6:00:00 PM	7:30:00 PM	SHchildABR2	Sensible Heat ChildA Bedroom # 2			
6:01:00 PM	6:06:00 PM	LightMastBath	Master Bathroom Lights			
6:02:00 PM	6:06:00 PM	Light1stBath	1st Bathroom Lights			
6:02:00 PM	7:02:00 PM	CookTops	Cooktops			408.04
6:05:00 PM	6:06:00 PM	SinkMastBathH	Sink event - Master Bathroom hot water	2.38	0.63	
6:07:00 PM	6:30:00 PM	LightMastBed	Master Bedroom Lights			
6:08:00 PM	6:18:00 PM	Light2ndBath	2nd Bathroom Lights			
6:09:00 PM		MicrowaveCLR	Clear Microwave (equivalent to opening and closing the door)			
6:10:00 PM	6:25:00 PM	ShowerMastBathH	Shower even - Master Bathroom hot water	53.00	14.00	
6:10:00 PM	10:40:00 PM	LivingRATV	Living room - Active TV			
6:10:00 PM		MicrowaveON	Turn Microwave ON			
6:20:00 PM	12:00:00 AM	Light2ndBed	2nd Bedroom Lights			
6:26:00 PM	6:27:00 PM	SinkMastBathH	Sink event - Master Bathroom hot water	2.38	0.63	
6:30:00 PM	7:30:00 PM	SHchildBBR3	Sensible Heat ChildB Bedroom # 3			
6:30:00 PM	8:15:00 PM	ComputerACB	Bedroom # 3 - Laptop			
6:30:00 PM	8:30:00 PM	Light3rdBed	3rd Bedroom Lights			
6:30:00 PM	10:00:00 PM	ComputerACA	Bedroom # 2 - Laptop			
6:30:00 PM	10:30:00 PM	SHparentAK	Sensible Heat ParentA Kitchen			
6:40:00 PM	6:41:00 PM	KitchenSinkH	Kitchen Sink hot water	2.38	0.63	
6:40:00 PM	6:47:04 PM	CanOpener	Can Opener			
6:40:00 PM	6:50:00 PM	HandMixer	Hand Mixer			5.50
6:55:00 PM	6:56:00 PM	KitchenSinkH	Kitchen Sink hot water	2.38	0.63	
7:00:00 PM	7:15:00 PM	KitchenLights	Kitchen Lights			
7:00:00 PM	7:15:00 PM	LivingRLights	Living room Lights			

7:05:00 PM	7:06:00 PM	KitchenSinkH	Kitchen Sink hot water	2.38	0.63	
7:15:00 PM	8:30:00 PM	DiningRLights	Dining Room Lights			
7:30:00 PM	8:15:00 PM	KitchenLights	Kitchen Lights			
7:30:00 PM	8:15:00 PM	SHchildALR	Sensible Heat ChildA Living Room			
7:30:00 PM	8:15:00 PM	SHchildBDR	Sensible Heat ChildB Dining Room			
7:30:00 PM	10:30:00 PM	LivingRLights	Living room Lights			
8:15:00 PM	8:16:00 PM	KitchenSinkH	Kitchen Sink hot water	2.38	0.63	
8:15:00 PM	8:25:00 PM	Light2ndBath	2nd Bathroom Lights			
8:15:00 PM	11:59:00 PM	SHchildABR2	Sensible Heat ChildA Bedroom # 2			
8:15:00 PM	11:59:00 PM	SHchildBBR3	Sensible Heat ChildB Bedroom # 3			
8:20:00 PM	8:21:00 PM	SinkMastBathH	Sink event - Master Bathroom hot water	2.38	0.63	
8:28:00 PM		Dishwasher	Dishwasher			
8:30:00 PM	10:30:00 PM	KitchenLights	Kitchen Lights			
8:35:00 PM	8:36:00 PM	KitchenSinkH	Kitchen Sink hot water	2.38	0.63	
8:50:00 PM	8:51:00 PM	KitchenSinkH	Kitchen Sink hot water	2.38	0.63	
9:00:00 PM	9:01:00 PM	KitchenSinkH	Kitchen Sink hot water	2.38	0.63	
9:00:00 PM	9:15:00 PM	Light2ndBath	2nd Bathroom Lights			
9:05:00 PM	9:06:00 PM	SinkMastBathH	Sink event - Master Bathroom hot water	2.38	0.63	
9:30:00 PM	9:31:00 PM	KitchenSinkH	Kitchen Sink hot water	2.38	0.63	
9:40:00 PM	9:41:00 PM	KitchenSinkH	Kitchen Sink hot water	2.38	0.63	
9:45:00 PM	10:00:00 PM	Light2ndBath	2nd Bathroom Lights			
9:55:00 PM	9:56:00 PM	SinkMastBathH	Sink event - Master Bathroom hot water	2.38	0.63	
10:00:00 PM	10:15:00 PM	HeatingPad	Heating Pad			
10:00:00 PM	11:00:00 PM	LightMastBed	Master Bedroom Lights			
10:00:00 PM	11:30:00 PM	BlueRayM	Master Bedroom - BlueRay			
10:00:00 PM	11:30:00 PM	MastBedATV	Master Bedroom Active TV			
10:00:00 PM	11:59:00 PM	PortableFan	Portable Fan			
10:00:00 PM	11:59:00 PM	SHparentBMBR	Sensible Heat ParentB Master Bedroom			
10:05:00 PM	10:06:00 PM	KitchenSinkH	Kitchen Sink hot water	2.38	0.63	
10:05:00 PM	10:25:00 PM	LightMastBath	Master Bathroom Lights			
10:20:00 PM	10:21:00 PM	SinkMastBathH	Sink event - Master Bathroom hot water	2.38	0.63	
10:30:00 PM	10:40:00 PM	LightMastBath	Master Bathroom Lights			
10:30:00 PM	11:59:00 PM	SHparentAMBR	Sensible Heat ParentA Master Bedroom			
10:40:00 PM	10:41:00 PM	SinkMastBathH	Sink event - Master Bathroom hot water	2.38	0.63	

www.ingramcontent.com/pod-product-compliance
Lightning Source LLC
Chambersburg PA
CBHW081851170526
45167CB00007B/2966